HOW HEMLINES PREDICT THE ECONOMY

How Hemlines
Predict the Economy

Explanations, Rationalizations,
and Theories on Everything

PETER FITZSIMONS

Skyhorse Publishing

Skyhorse Publishing books may be purchased in bulk at special discounts for sales promotion, corporate gifts, fund-raising, or educational purposes. Special editions can also be created to specifications. For details, contact the Special Sales Department, Skyhorse Publishing, 555 Eighth Avenue, Suite 903, New York, NY 10018 or info@skyhorsepublishing.com.

www.skyhorsepublishing.com

10 9 8 7 6 5 4 3 2 1

Library of Congress Cataloging-in-Publication Data
FitzSimons, Peter.
How hemlines predict the economy : explanations, rationalizations, and theories on everything / Peter FitzSimons.
p. cm.
ISBN 978-1-60239-311-0 (alk. paper)
1. Australian wit and humor. I. Title.
PN6178.A8F57 2009
828'.91402--dc22
2008029913

Design by LeAnna Weller Smith
Printed in China

*Always the struggle of the human soul is to break
through the barriers
of silence and distance into companionship.
Friendship, lust, love,
art, religion—we rush into them pleading,
fighting, clamoring for the
touch of spirit laid against our spirit.*

E.B. WHITE

Contents

Introduction

What follows is a collection of theories, conundrums, observations, quotations, and whatnots I have collected since forever. The theories may be described as being of the folk variety—the kind that people repeat to one another at dinner parties and at the bar, as well as the sort of things they tend to think about as they wait for traffic lights to change—mixed together with a few others that are actually taught in college but are still of the "folk" variety.

My first *Theories of Life* book came out in 1991. In my travels since then, I've collected a few more theories and have gotten rid of those that have not passed the test of time. I have also peppered this edition with quotes from a few famous figures that are practically theories in themselves.

The wider theories come from pretty much everywhere. As you might appreciate with such a word-of-mouth book, it has not always been easy to identify who told me what, and what I heard where, but I have at least honored my original promise from all those years ago and have acknowledged direct contributors in the back of the book.

Some of these life theories genuinely have a scientific claim to accuracy, while the others would not deign to turn to something so hard, cold, and damned soulless as science to verify or prove their claims.

But how accurate is "accurate"? Well, I've heard it said that Einstein would not publish a theory unless he felt it was accurate 100 percent of the time, and Sigmund Freud was satisfied if his theories were correct 95 percent of the time.

My own cutoff point is . . . 51 percent.

I know this probably means that Einstein, Freud, and FitzSimons will not ultimately go down inextricably linked in history, but what do I care? If I can call heads or tails and be right 51 percent of the time, I'm bound to come out a winner sooner or later.

In all seriousness, though, I do claim that at least 51 percent of the theories are correct a lot more than 51 percent of the time, and it's just possible some of them might show you some discernible tracks in the otherwise impenetrable jungle of life.

Read 'em and weep, Sigmund. You too, Albie Einstein . . .

HOW HEMLINES PREDICT THE ECONOMY

CHAPTER 1

RELATIONSHIP THEORIES

SIDE-OF-THE-BED THEORY

All else being equal, the dominant partner in any sexual relationship will always sleep on the side of the bed nearest to the most likely source of danger.

So if the bedroom door is closer to the left side of the bed, the dominant partner will always sleep on that left side, and the opposite if the door is on the right.

Now, if the door opens to the foot of the bed and the most likely source of danger is thus not on the flanks, then look to the windows . . .

In my own experience, this theory is almost infallible when the direction of the most likely source of danger is as clear as whom the dominant partner is. The only exception seems to be when there are young children in the family—in which case the dominant partner ensures that the submissive partner will sleep closer to the wailers.

If the theory works for you, I take a bow. If it doesn't, re-examine your first premise—are you really the dominant partner? No, Mr. Mitty, I really didn't think so.

Another theory along these lines maintains that sexual partners always lie on the side of the bed that corresponds to the side of the car where they would normally sit when driving together, with the habitual driver on the left side of the bed and on the other side in the right-hand-driving countries. I prefer the first theory, not only because it includes a lot more sex and violence, but, more important, because I find it is correct far more often.

HAND-HOLDING

Of course, the problem with the preceding theory is that while it may give couples themselves a better clue about who is winning the war of the sexes, the rest of us will have no clue whom the dominant partner is simply by glancing at them, unless we have access to their bedrooms at midnight. But there is another way . . .

Another theory holds that for two right-handed people, the dominant partner will always be on the left side when they walk along holding hands, thus using his or her marginally more comfortable right hand to hold with.

If a woman has to choose between catching a fly ball and saving an infant's life, she will choose to save the infant's life without even considering if there are men on base.

—Dave Barry

THE BEDROOM CEILING

Women will always notice that the bedroom ceiling needs painting before their husbands do.

Housepainters—who are in the perfect position to know who thinks what should be done in couples' houses—maintain that this is absolutely gospel. It is one of the great mysteries of our time as to why this should be so, but in the housepainting fraternity it is a well-documented phenomenon.

TILTED-HEAD KISSING

British, Australian, and New Zealand lovers generally kiss with their heads tilted to the right, while American and Canadian lovers tilt their heads to the left (and the French and Italians have a bet each way, twenty times a day).

I wish I could claim originality for this little earth-shaker, but I can't; it came to me from a friend who has long noted the phenomenon. But for my own part I can verify its truth, based on those oh-so-rare occasions I got lucky while living in Ohio in the late seventies.

Americans do kiss with their heads tilted to the left, which can be verified by watching American and Canadian movies and soap operas. The reason for this is obvious once it is stated: Most first kisses are performed in cars, and from a physical point of view it is natural that those who drive left-hand-drive vehicles should incline their heads to the left (think about it now) when they kiss somebody on their right. The reverse of course applies in those countries where the steering wheel is on the right side.

Once this pattern has been established in the habits of kissing participants, it also becomes the norm for situations outside cars. These habits then continue outside the car as a matter of course . . .

Brilliant!

No, of course it doesn't work 100 percent of the time, but we're also well above the realms of the ol' 51 percent for all that . . .

If you want something said, ask a man; if you want something done, ask a woman.

—Margaret Thatcher

4

THE EYES HAVE IT

This one sounds stupid, looks stupid, smells stupid, and feels stupid, and normally I would reject it as in fact being stupid. But on a whim, I checked it out and found it remarkably accurate. The theory is that men are happiest living with women whose eyes are the opposite in general color to those of their mother. If a man's mother's eyes are dark, he will marry a light-eyed woman. If she is a light-eyed woman, he will marry a dark-eyed woman.

YOU ARE GETTING SLEEPY

The reason why men get very sleepy after an orgasm, while women often perk up, is apparently this . . .

Evolution has deemed it such that, generally, the male will be endowed with the aggression necessary to hunt down the female and have his wicked way with her . . .

Of course, we humans are now more or less civilized, and aggression should have no part in the pursuit of one's sexual urges, but in the animal kingdom—whose laws we often still obey in spite of ourselves—this instinctive aggression is essential for procreation. After the sexual act is performed, the aggression of the male, which has powered him up to this point, can now become a liability for her and it is necessary to knock him out so the female can make her escape.

By all means marry; if you get a good wife, you'll be happy.
If you get a bad one, you'll become a philosopher.

—Socrates, Greek philosopher (469–399 B.C.)

THE JELLY-BEAN THEORY OF SEX

The author does *not* subscribe to this theory, mind you, but it is too much fun to leave out of the book. Of unknown origin (though I suspect it emerged from somewhere in Australia), the theory starts with a new couple putting a single jelly bean in a jar every time they have sex. Once married, they must continue to do so for the first year. After twelve months of wedded bliss, however, they must take *out* a jelly bean every time they have sex.

If followed faithfully, the theory has it that they will never reach the bottom of the jar because the enthusiastic frequency of that first flush of a new relationship will always be greater than the frequency for the rest of the relationship.

One of my theories is that men love with their eyes; women love with their ears.

—Zsa Zsa Gabor

If all the girls who attended the Harvard-Yale game were laid end to end . . . I wouldn't be at all surprised . . .

—Dorothy Parker

CHAPTER 2

CONUNDRUMS

What follows are, if I do say so myself, some of the truly great questions of our time . . .

SIMPLE QUESTIONS

Why do they put expiration dates on cartons of sour cream?

How does medicine know where to go?

How come mosquitoes sound louder in the dark?

Why is the fluff that catches in your belly button always blue?

What did Jehovah see, anyway?

PRESSING QUESTION

At pedestrian crossings and elevators, why do people frantically press the button in the manner of laboratory rats pressing levers in experiments in the hope of getting a piece of cheese? Do they really believe that the sensors are rigged so that ten presses in five seconds will program the light to change or the elevator to come sooner?

MOVIE QUERIES

In movies, why do actors who get hung up on immediately get a dial tone, when this never happens in real life?

And speaking of movies, how come in every film you ever saw, the good guy firing shots at fifty bad guys always makes his shot, while for the life of them, the fifty bad guys can't bring off more than minor flesh wounds on the good guy? The rough equivalent of this in martial arts fights in movies is that for some reason the bad guys are always kind enough to attack the good guy one at a time instead of rushing him all at once.

How is it that when villains fire bullets at Superman, the bullets bounce off, and yet when they then throw the empty gun at him . . . he ducks?

HOME TRUTHS

Why do people run over a piece of string a dozen times with their vacuum cleaner, reach down, pick it up, examine it, then put it down to give the vacuum one more chance?

Why is it that if you actually look like your passport photo, then you are obviously too sick to travel?

Why is it that when you work out how much your renovations could possibly cost, the final bill is always exactly double?

RAILWAY GATES

Why, all over the world, does the railway gate not lift until a good thirty seconds after the train has passed through?

Why? What atavistic urge within us makes the barriers remain down long after all possible danger has passed?

As a child, I have a very clear memory of once being forced to retrieve a baseball from between the railroad tracks. Just as I arrived on the embankment, a huge freight train appeared and *whooshed* . . . over my baseball. Of course, I stood and watched in great fear until it had passed, and it was a good thirty seconds before I had the courage to quickly dart out, get it, and scurry back . . . even though I could see for a mile in each direction that the track was quite clear.

What interests me is how the adults who program the gates in all the many countries to which I have been all surrender to the same built-in fear, even when, logically, they must appreciate that it's quite safe to cross the instant the train has passed.

RELIGIOUS VISIONS

How come the most miserable-looking people you ever saw are those who claim to have found the joy of knowing Christ?

And how is it that the visions of Jesus Christ and the Virgin Mary and so on occur only in Christian countries, while the visits from Allah are heard of only in Islamic countries? It's really so convenient it's incredible. Imagine the confusion if Jesus Christ chose to make an appearance in someone's cereal in Mecca or Muhammad himself turned up in someone's oatmeal in Dublin. On the other hand, are their images appearing and just being wolfed down merely because they are not being recognized?

MAYBE MEN

Why do so many people with the initials MM rise to positions of rather enigmatic guruness in various fields?

Just for starters, there are Marshall McLuhan, Malcolm McLaren, and Mickey Mouse, not necessarily in order of importance.

PASTE

How can it be that every paste manufacturer you ever heard of—Elmer's, Avery, etc.—is so totally incapable of constructing a paste brush that will reach the bottom of the container?

MAKING LOVE

How is it that there is a very strongly held conviction out there that men generally make love all the time, always have affairs on the side, and are generally infinitely the more horny of the two sexes . . . when in fact (in the heterosexual world, at least), the sum total of acts of intercourse committed by men since the beginning of time, is *exactly* equal to the number committed by women?

We're not talking more or less equal here, or give-or-take-one equal . . . we're talking right on the pupil of the bull's-eye equal. So exactly equal that no other equation in the history of the world even comes close for equalness.

THE PARADOX OF OUR AGE

We have bigger houses but smaller families;
more conveniences but less time.
We have more degrees but less sense;
more knowledge but less judgment;
more experts but more problems;
more medicines but less healthiness.
We've been all the way to the moon and back,
but have trouble in crossing the street to meet our new
neighbor.
We built more computers to hold more copies than ever,
but have less real communication.
We have become long on quantity,
but short on quality.
These are the times of fast foods and slow digestion;
tall men but short characters;
steep profits but shallow relationships.
It's a time when there is too much in the window
but nothing in the room.

—The 14th Dalai Lama

THE GREAT BOXING CONUNDRUM

How can it be that every boxer you've ever heard of—I mean *every* boxer you've ever heard of—always has a for-and-against record that is flattering to him, with always more victories than losses?

Even on the lowliest undercards of the most abysmal fights, the two boxers always have more victories than losses to their credit. In the red corner ... Johnny Pug with a record of 16–11, 4 by knockout, and in the blue corner ... Sammy My-Nothe-Ith-Broke, currently on 19–4.

As to the champions, they always have records like 61–2 and 38–4. The only possible explanation for this is that there must be five or six boxers out there, as yet unheard of, who currently have records of 2–614 and 3–947.

IT ALL COMES OUT IN THE WASH

How is it that when you mix clothing of different colors in the washing machine, the colors start to run (as in the red from your red shirt runs into your white shirt), but when you wash the white shirt again on its own, nothing—and I mean nothing—will convince the red that it's time to leave your white shirt? Why is it so, Julius?

YOU SHOULD BE SO LUCKY

Why is it so that in the whole history of the world, no one has ever marked a lottery card 1, 2, 3, 4, 5, 6? That sequence of numbers is just as likely to occur as any other group of numbers—say, 3, 5, 19, 21, 28, 30—yet the 1–6 sequence has the added advantage that if it does come up, you will be the only person to have marked the Lotto card as such and thus won't have to share the prize.

CHAPTER 3

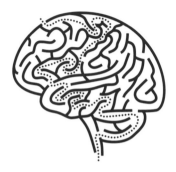

PSEUDO-PSCIENTIFIC

Pseudo-pscientific is as close to science as this book gets. These are the sort of theories people often repeat to one another with the sort of breathless air of ". . . and it's actually true." In fact, they are all at least 51 percent true.

YOUR LYING EYES

The theory goes that if right-handed people are asked a searching question—"Darling, where did this receipt for a room at the Midnight Love Motel come from?"—and they wish to lie, then they will have to rely on the left hemisphere of their brains, where imagination resides.

Consequently, they will generally look to the right while pondering their reply. If they wish to tell the truth, however—"Oh, that? Let me see . . . I think that was the day I first started seeing Molly"— they will delve into the right side of their brains, where memory resides, and thus will look out to the left.

The reverse is true for left-handed people. If they are lying, they will look out to the left.

CRAPPY NAPPIES

People who constantly had dirty diapers as children are the ones who are now pathologically untidy.

(Sincere apologies to Sigmund Freud on this one, because I think it was originally his theory.) The theory goes that if as a baby your diaper was changed twenty times a day (as in immediately after it got wet or soiled) or if as soon as you got a little dirt on you, or some ice cream on your cheek, your mother or father quickly washed it off, then your tolerance for dirt and untidiness will be very low. Ergo, for the rest of your life you will be physically, mentally, and emotionally uncomfortable when confronted with dirt or untidiness. You're known as anal-retentive and you are the sort of people who wash their cars three times a week, make your beds in the morning, actually do the dishes right after dinner, and just generally drive us anally expletive people *nuts.*

We anally expletive people, on the other hand, are the ones (often the youngest in large families, as I was) whose mothers never had time to get to their diapers right away, nor even had the inclination to wipe away that slather of ice cream spread all over our cheeks and all down our fronts. Our threshold of tolerance for dirt and untidiness therefore came to be as tall as the Eiffel Tower and arguably is worthy of similar awe.

Personally, I bow to no man in my ability to wear dirty socks for months on end and, to be frank the only reason I ever change my socks at all is not because I have any personal desire to be clean so much as because of reluctant deference to the far more anal-retentive sensibilities of my fellow man and—more particularly—woman.

A CHILD'S HEIGHT

Should you want to know what height a child will be when it is an adult, take its height at age two (not a day before or a day after) and double it.

I was born and brought up on this one and in my own case it was accurate to the nearest millimeter. Even the highbrow doctors generally concede the correctness of this theory.

YOUR CHILD'S SEX

You can determine the sex of your child by timing intercourse precisely.

For a girl, couples should have intercourse two or three days before ovulation. For a boy, couples should have intercourse on the day of ovulation or soon after.

An alternative theory has it that particularly frequent sex leads to boys, while occasional sex leads to girls. If true, it would help explain the standard chestnut that nature compensates for the loss of men during wars by creating a preponderance of boy babies right after a war. It would also help explain the oft-observed phenomenon that in large families the boys tend to be among the first born while the girls are more often born later.

PREGNANCY

Meantime, an old folk theory (of origins unknown) has it that there is one surefire way to determine the sex of a child before it is born. It requires the mother-to-be to hand over a hair from her head and her wedding ring. The hair is tied to the ring. Now the woman lies on the floor as the ring is suspended over her belly. The theory has it that if the ring swings around in circles, then the baby is a boy, while if it swings back and forth sideways, the baby is a girl.

TESTOSTERONE AND DIGIT RATIO

The greater the ratio between the length of your ring finger and your index finger, the more testosterone you were exposed to in the womb.

Hold up your hand. If you're a man, your index and your ring finger will be close to the same length. If you're a woman, the ring finger will be clearly smaller than your index. As a rough rule of, er, thumb, the more masculine the man, the longer that ring finger.

WELL, I BELIEVE HIM

An orthopedic surgeon wrote to me about this one and he positively swears it is true. I quote him as follows:

> One truism of orthopedic surgery includes the X-ray appearance of fractures in the male hips. When one looks at the X-ray of a pelvis and there is a fracture in one of the hips, one invariably notices that the penis is pointing toward that side of the pelvis.
> Thus the truism of orthopedics in relationship to fractures is that the penis always points to the side of injury—to the weakest hip.

This man has performed thousands of hip operations over the past twenty years, and he says if there is one thing in his life he is certain of, it is the above dictum.

*If A is a success in life, then A = x + y + z. Work is x,
y is play, and z is keeping your mouth shut.*

—Albert Einstein

CATERING

The real reason humans stopped being hunter-gatherers and settled down to cultivate fields was because they discovered alcohol.

See, alcohol is often made from grains, and if you wanted to brew and distill alcohol in any large amounts, you needed to settle down to give it time to really achieve its potential, and to develop the containers that could house your precious drop. You couldn't lug large quantities of beverage around with you, so if you wanted greater quantities, you needed to stay put and start to propagate the stuff you could make alcohol from.

The Sumerians of the Tigris and Euphrates river basin—the "Cradle of Civilization"— started it all. Tablets covered with cuneiform script show a form of beer being drunk. The Babylonians continued the obsession, brewing over twenty types of beer.

Alcohol became key to social interactions and traditions; many festivals centered on some form of ritual imbibing! Some researchers have even suggested that the relaxing side effects of alcohol were what enabled people to live together so closely in such large numbers.

20

Of course, where you could obtain your liquor from a mobile source, for example, koumiss, the fermented mare's milk still drunk in Mongolia, there really wasn't the same pressure to settle down . . .

JET LAG

Jet lag always hits you more as you fly from west to east, against the direction of the sun, than when you fly from east to west.

Ask anybody who does the New York/London trip regularly. Generally, when you arrive in London you are totally exhausted for at least a week. But when you fly to N.Y. you're ready to rock 'n' roll shortly after landing. The people who fly the route regularly note that going west is more like shifting to weekend time, when you stay out late and wake up to a late brunch. Going east, on the other hand, is as if you're trying to go to sleep at 7 P.M. and are then awakened at three in the morning.

According to David Neubauer, who is the world's expert on the subject, "It is people with rigid internal clocks who usually suffer the most from jet lag." Early birds, people who feel chipper in the morning and are asleep by 10 P.M., are at high risk for jet lag. Night owls, people who have no trouble staying up late and sleeping late, tend to be more flexible with body-clock adjustments.

CHAPTER 4

SPORTING THEORIES

SOCCER PLAYERS ON A BENCH

Ten soccer players on a ten-yard bench will instinctively arrange themselves so that each player will have exactly one yard of space to himself and no one will be touching the other. Ten *French* soccer players on a ten-yard bench, however, are just as likely to be sitting beside each other within a space of six yards.

In fact, they will sit precisely as they have landed, and the draping of one player's hand over another player's thigh, for example, will raise no cry of objection.

Trust me on this one. If I spent four years playing rugby in France for one thing, it was to postulate and then confirm the truth of this theory. In fact, of all the theories gathered in this collection, this is the one I judge to be closest to being rinky-dink, ridgy-didge, fair dinkum, 100 percent correct.

The same kind of rule applies incidentally to automatic teller machines. In most Anglo countries, people will give themselves a good yard's distance between each other, while in France they're just as likely to be pressed up against you, and in Tokyo, you actually have to apply a little backward pressure on the line behind you to avoid smudging the screen with an impression of your lips.

COACHES

It is a sporting chestnut, but this theory has it that there are only two types of coaches: those who have been fired, and those who will be fired. It just about never fails! The political equivalent was most succinctly uttered by well-known British figure Enoch Powell when he famously said, "All political careers, unless they are cut off in midstream at a happy juncture, end in failure . . ."

BETTING

I have worked in the *Sydney Morning Herald* sports department for twenty years. Unlike me, my colleagues are serious sports writers who spend a lot of their working day attending practices and games, and talking between times to players, coaches, and administrators about the teams they cover. Each Friday, based on their accumulated knowledge of the previous seven days, they make predictions in the *Herald's* published bets about which team will win.

But here's the thing: In the mid-1990s our best bettor won with an accuracy rate of 67 percent and was congratulated for doing so. One of my colleagues, however, noted that in that particular year, if you simply picked the home team to win every time, you would have won with an accuracy record of 69 percent! It revealed a truth that I have embraced ever since, and despite limited inside knowledge of what is happening week by week with the various sports teams, on the years I have ruthlessly applied the notion that the home team will most likely win (unless completely outclassed by a team much higher on the ladder), I have done extremely well in betting and, on occasion, have even won.

THE LAST ONE OUT

When any two football teams run out onto the field, the last player out will almost always be either the most superstitious of the team or the most paternal, or a combination thereof.

Tap, tap, tap. The knock comes on the door. It's time, guys.

The word comes to the locker room at about noon. There is a sort of collective "oooooommphhhh" released from everyone, as the battle cry sounds. The captain, with ball in hand, strides purposefully to the door as the rest unwittingly fall into line.

Immediately behind the captain will be the cocaptain or, if no cocaptain has been designated, the second in line will frequently be the guy who thinks he should be cocaptain and wishes to create the impression that he *is*, to the crowd, at least. Then comes the bulk of the rest, lining up every which way . . . and at the very back, an interesting little scene is probably being played out.

Those guys who are superstitious naturally tend to compete for the back of the line because it is so obvious. I mean, they're not going to be superstitious for the eleventh or the seventh spot, are they? With the battle cry still ringing, no one ever thinks to actually count out their position in the line and ensure that they've got the right one. But the *very back* of the line, that is another matter. For a lot of guys, the back of the line provides a little anchor to cling to in the flood of tension that besets everyone—and can even give strength once secured. Usually, the superstitious one will also be one of the less experienced players because superstitions like these tend to fall away as the players get older.

One of the other sharks circling around up the back might be the paternal figure who didn't make it to captain. Though not the official shot-caller, he finds something comforting in ushering his charges out before him in the manner of a mother hen with her chicks

Who wins this weekly battle? It depends. But the bottom line is that no one ever ran out last onto the football field merely because that's the way the cards fell.

SPORTS TOURS

When on a sports tour, the average mental age of the group drops about ten years, although as the group gets older, the age-drop gets larger.

I write this as one who has been on many tours and was as much an observer of the phenomenon as a victim/celebrant of it. Without going into details— mercifully—when I was on tour for the first time, as a fifteen-year-old, my teammates and I behaved like five-year-olds. At the age of twenty-five, we could have held our own in fart jokes with a bunch of fifteen-year-olds, and even at thirty-one, we could find high hilarity in things that might have made normal twenty-one-year-olds blush with the infantilism of it all. Don't know why that was, but it just was—and still is.

I can't resist finishing with my favorite rugby tour yarn, which I think demonstrates that the same rule still applies, even when the tourists are much older.

A few years back, the old and bold Hunters Hill Cats Golden Oldies rugby team was on a flight from Sydney, Australia, for some rugby revelry on the Gold Coast. The fact that the airline had not spotted them as a rugby team was likely due to the fact that they had booked the flight as an all-female bowling league and therefore were all seated together. Mistake. Really big mistake.

The flight attendants that day were headed by the truly gorgeous Brenda . . .

And as the drinks cart did its job and the Golden Oldies focused progressively on her lovely form, they soon forgot their baldness, creaking joints, and obesity and were taken back to their youth, back as far as 1977, when they had gone through the entire season undefeated. The more they gazed at Brenda, the more they remembered, until, just before the captain announced that the drinks service was mercifully finished, it happened . . .

From the midst of this drooling mob came: "Hey, Brenda, what's ya muvva's phone number?"

THE RAREST THING TO SEE IN SPORTS IS . . .

. . . a soccer goalie who has a goal scored against him and accepts it with a modicum of grace.

The next time you see highlights of a soccer match on TV, from anywhere in the world, take a look at the reaction of the goalie immediately after the ball whooshes past his sacred person and into the net.

Every time a goal is scored—I mean *every time* a goal is scored—the goalie's first reaction will always be to glare at his own two defenders and to imply that the fact that his goal was breached was most definitely their fault and most certainly not his (heaven forbid!).

It doesn't matter how brilliant the goal, how diligent the defense, the goalie will always come up glowering, pouting, hands belligerently on hips, to glare at the low dogs who dare call themselves his teammates . . . and generally do everything possible to lay the blame elsewhere.

WHY THERE ARE ALWAYS RIOTS AT SOCCER MATCHES

I think it was Desmond Morris who first expounded this theory but, like *The Naked Ape*, I'm going to appropriate some of it and elaborate upon it. The reason that there are so frequently riots at soccer matches, as opposed to *full-on contact* football games is this . . .

All types of football games are extensions of the tribal instinct. The players of a team are like warriors of a tribe, fighting the warriors of enemy tribes . . . In football, the rules allow the warriors to come into physical contact. Players tackle each other and wrestle each other to the ground. Thus, vicariously, the spectator is also in combat; there is an outlet for their emotions.

But in soccer, the spectator is frustrated, as there is no contact allowed between the warriors. Furthermore, in football, points are scored often enough to provide some sort of emotional release one way or another; in a soccer match there is no such frequent outlet for the emotions. Often the result is a very frustrating draw, and in a whole match there might be only one real moment of exaltation when your team scores. Put it all together for soccer and you have a bomb waiting to go off. And sometimes it does.

A TEAM'S BACKBONE

As the game of football progresses, there is one surefire way to judge the backbone left in the team for the fight: In a shoulder-to-shoulder huddle, with each team member putting in his two cents' worth, it is sure that they want to go on with it and strike back at the opposition. If they stand shoulder to shoulder to shoulder gazing at one another, then this is less likely, and if they stand with a good yard between each of them, then the game is lost and they know it and don't care for the struggle anymore.

WHO WON?

All over the world, sports fans, when talking about the performance of a team of any sport, will either say *we* won, or *they* lost.

THE GUARDIAN-IVY COMPLEX

There is a syndrome in Australian life that is far more prevalent than the famed Tall Poppy Syndrome (which is a famed *national* belief that holds that once someone has achieved a certain celebrity status, there will be a natural tendency for the media and certain members of the public to want to cut them down to size).

I call the lesser known syndrome the Guardian-Ivy Complex. That is, as soon as anyone takes a shot at a Tall Poppy—most particularly in the fields of sports and entertainment—myriad media guardians will start swarming to the Poppy's defense, wrapping themselves around the trunk of the Poppy to take the evil blows themselves. Self-conscious in their saintliness, these guardians are there to defend to the death the right of the Poppies to continue blossoming untouched. You can call them old-fashioned if you like, but they just want to see the Poppies get a fair try!

THIRD-PERSON THEORY

The surest sign that a celebrity is getting lost in their own stardom is when they begin to refer to themselves in the third person.

As a young man, Fred Jones Jr. dreamed of winning an Olympic gold medal in running. Blessed with talent, powered by ambition, he works hard at it and rises through the ranks to the point that both the public and the media begin to pay attention to him. Pretty soon, his name starts to appear in the local and even national media.

FRED JONES JR. WINS AGAIN
FRED JONES JR. RAISES THE BAR
FRED JONES JR. SETS NEW RACE RECORD IN
HIS AGE GROUP!

Inevitably, everywhere Fred goes now, people are starting to point him out, whisper behind their hands, even approach him for autographs. It is a constant refrain, at the field, on the bus, at school, in the papers: "Fred Jones Jr. . . ." "There goes Fred Jones Jr. . . ." "I think that's Fred Jones Jr. . . ."

It is intoxicating, seductive . . . silk for the soul.

As the great day comes when he is actually selected to represent America at the Olympic Games, then what has gone on before is nothing compared to what happens now.

FRED JONES JR. TO GO TO BEIJING
FRED JONES JR. INCLUDED ON RELAY TEAM
FRED JONES! FRED JONES! FRED JONES!

Pretty soon, the inevitable happens. The sound of his own name is so sweet, so seductive, that the day comes when Fred can't resist trying it out himself. So instead of saying, after winning a race, "I had a great day out there," he says, "It was a great day for Fred Jones Jr." Instead of saying, "I am not sure which product I will now endorse," he says, "Fred Jones will do what's best for Fred Jones."

And so it goes . . .

SUNGLASSES

And the second surest sign that athletes are getting lost in their own celebrity is when they start to wear sunglasses, even on dank, dark days!

Not so long ago I was walking between the Sydney Football Stadium and the Sydney Cricket Ground, beside the practice nets, when I stopped long enough to observe the New South Wales Cricket team practicing, two days prior to a game. Actually, it was doubtful that the game would be able to take place at all, as Sydney was going through a notably rainy period. But, despite the inclement weather, five of the players wore sunglasses. I looked closer. Every one of them had also represented the national team and so they knew what it was to be recognized, to be awed at, and so presumably had also earned the right to wear sunglasses on cloudy days. Mick Jagger, eat your heart out.

Further back still, I was in South Africa for the 1995 Rugby World Cup and was chatting to captain Sean Fitzpatrick on the night of the World Cup final, where his team was due to play the Springboks. The subject of the Wallabies' poor performance came up, and I will never forget Sean's words.

"I knew they weren't going to do anything," he said, "from the moment I was watching on television and saw them get off the bus for their first match."

"What on earth do you mean?" I inquired. "How could you possibly know that?"

"Oh, you know," he replied breezily, "the sunglasses . . . the attitude."

And it was true. That generation of Wallabies really did have something rock-star about them, and had rather played like that.

PHOTOS . . .

Meanwhile, the surest sign that the media itself has fallen in love with a particular sports personality is when it starts to publish photos of the star from childhood years.

When an athlete is selected to a U.S. Olympic team, or any team of athletic stature, it is all admirable enough . . . but it does not warrant in itself an exploration or illustration of the star's childhood. But now, let them win an Olympic gold medal! At this point they are immediately in sports icon territory, living the dream that so many of us had when we were children . . . and so it immediately becomes fascinating to know what the star looked like when they were children.

CHAPTER 5

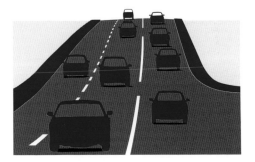

TRAFFIC LORE

RED CARS

In traffic, red cars love to beetle along the road together in packs. Brown cars are not nearly as gregarious, but they like nothing better than to hang just off the edge of a pack of red cars and hare along thus.

The authority for this comes from those taxi drivers who have been on the road for the past forty years and have long observed such behavior. They claim, in all seriousness, the truthfulness of this theory.

They get support in their contention from other roadsters like bus drivers and couriers. Honest they do. There is also an associated theory to the effect that while white cars care naught for the company of any other cars on the road other than what chance delivers, the fact is that when parked, they do just adore to park beside other white cars.

PARKING-LOT MAGNETISM

New arrivals in parking lots invariably succumb to the very strong magnetic effect of lone cars parked in open spaces.

Though I've no doubt that there are a few doubting men and women left over from the last theory, they should not doubt the truth of this one. It is another one that gets in the high nineties for accuracy, if admittedly in the very low tens for relevancy to the Great Questions of Life. But what do I care? It interests me.

Longtime parking-lot attendants have noted the phenomenon. All else being equal, the cars that have arrived later love nothing better than to sidle right up alongside the loner and snuggle down there for a spell until their owners get back.

THOSE ITALIANS

Italians don't park, they stop. In Italy there is no notion of getting the car off the street so that other cars may pass by. Though parking a car usually involves the whole twisted process of turning the steering wheel, changing the gears, etc., and then performing all the convolutions of the classic parallel park with a reverse maneuver in a ninety-degree angle, degree of difficulty: 2.1 . . . in Italy they just turn off the ignition. So much easier that way, *non, amico mio?*

Although this rule holds true for Italians generally, it is not unfair to say that on many occasions the natives of Rovigo (where I once lived) neither parked nor stopped, so much as . . . crashed. It was even less trouble than formally stopping, even if it was hell on the paintwork.

As to the French, they seem neither to park, nor stop, nor crash. They simply abandon their cars at the closest point to where they need to go and let the devil take the hindmost, together with the subsequent traffic jams.

STRANGE BUT TRUE

If you've been lucky enough to see it, you know that the only humans who still negotiate rush hour traffic in brown Ford Pintos are packs of nuns.

Where do these nuns come from? And where are they going? Most important, what is their curious purpose on this planet that it involves going out only in rush hour traffic and only in brown Pintos? I know not, nor care much, but I know the phenomenon exists.

It's easy to miss if you're not aware of it, but now that I've clued you in, look for them and they will be there. Everywhere and always. Penguining their way through the traffic. Could it possibly be part of a strange new order: Sisters of Mercy Who Get Around in Brown Pintos in the Middle of Rush Hour Traffic?

I am not making this up.

SIDE OF THE SIDEWALK

In countries all over the world, people walk along sidewalks and hallways on the same side as they drive on the streets.

People drive on the right side in America and keep to the right when walking along Madison Avenue in New York, and Bourbon Street in New Orleans, and indeed all over. In France, they walk on the right along the Champs-Élysées, just as they drive on the roads, and on Via Veneto in Rome they charge along more or less on the same side as they drive on the Italian roads. (Actually, the Italians are frequently known to also drive their cars on the footpath, but that's another story.)

Now, check out those people charging up Madison and Bourbon Street on the left side of the sidewalk and making such heavy weather of it. They are not one of us—and are sure to be tourists or newly arrived migrants from left-side-drive countries.

Incidentally, this theory also holds true for people swimming laps in pools. In Australia and England they keep to the left as they do their laps; in France and America they keep to the right.

INDIAN ROADS

Then there is the rule that applies on single-lane Indian roads. Though ostensibly India is a drive-on-the-left country, the real rule that applies on single-lane roads is the same as the one that applies on their crowded pavements: The larger goes through.

When two vehicles meet on a single lane, each driver makes a subconscious decision as to who would come off worse in a head-on crash and the more likely loser gives way. I first formulated and tested this theory on one-thousand–odd miles of Indian back roads in the mid-1980s and it *still* works like a charm every time.

A similar rule of thumb applies to pedestrians on crowded Indian sidewalks, where big, beefy westerners reign supreme—so supreme that on pedestrian crossings they have even been known to try their luck against small vehicles—with mixed success.

PERSONAL SPACE

The sense that different nationalities have of personal space also extends to their vehicles.

Just as we generally feel uncomfortable if someone talks to us with their head inside that 1-foot radius that we regard as our personal space, we also have a good 2-yard buffer zone around our cars. When driving toward oncoming traffic, we tend to get a little panicky if an oncoming car comes within that zone and passes, say, at a distance of 1.5 yards.

Similarly, while the French feel uncomfortable if a conversing head comes within half a yard, the zone of comfort around their cars is commensurately only 1 yard.

By the time you get to the Italians and, God forbid, the Indian taxi drivers (most particularly of Mumbai), an oncoming car practically has to peel paint before their hearts skip a beat.

HATS

People with hats drive, on average, 20 mph slower than bareheaded drivers.

This is standard wisdom among taxi drivers, and it is uncannily true. For every passenger with a hat, deduct another 6 mph—or 9 mph if the hat is white. Further deduct 3 mph for every pipe smoker in the car. If they're towing a trailer, with a hat, two bales of hay, and a cattle dog, don't bother with any of these calculations, just pull over and read a book—it'll be quicker in the long run.

The really infuriating thing about these hat wearers is that if ever you get to talk to them, they can so smugly claim to have never had an accident in forty years of driving, though they've seen thousands, invariably via the rearview mirror.

YET MORE LEFT-HAND-DRIVE DOGMA

This is a theory about why the Americans, French, and others drive on the right, and the English drive on the left.

The reason why the English originally began to pass each other on the left is that that was the way the English knights in armor used to do it. Holding the lances in their right hands, they would pass each other at full gallop on the left. Because these august gentlemen were highly regarded in that nation, the custom of passing each other on the left passed into general usage. Which explains the English . . .

The French, being the French, insisted on doing it in the opposite fashion of the *Anglais* and passed each other on the right. The Americans, who when these things were being decided were in a similarly anti-English mood, started passing each other on the right and it proceeded to spread from there.

FOG

People who drive with their fog lights on when there clearly is no fog wouldn't have the foggiest about whether there is or is not fog and should clearly be deprived of their driving licenses for being fogwits and a danger to travel.

CHAPTER 6

MISHMASH

EVEN NUMBERS ARE THE GOOD GUYS

Most people, in their heart of hearts, have this ever-so-slight thing about even and odd numbers, believing the even numbers to be good, kind, honest, upright, and sensitive, and the odd numbers to be dark, nasty, and suspicious.

Think about it: How often when you must set your stove burner to a number between 1 and 10 do you set it on an even number and how rarely on an odd number? Isn't there something inherently nice about even numbers?

Was there ever such a suspicious, nasty, and horrible number as a 7, for example? Just look at the wretch. What goes into 7 but 1 and 7? Nothing. And why would they? Seven is a leper and an outcast and rightfully so. It's awful.

But then look at 8. Never a more honest and upright number walked the earth, an example to us all. Eight is a beautiful, symmetrical number made up of 4 × 2, themselves lovely numbers. 4 is particularly nice because it is made up of 2 × 2 and 2 is of course the loveliest of all numbers.

Nine is not as good as 8, but it is at least more personable than 7, because 9 is 3 × 3 and odd numbers can sometimes move in polite society if they are a square of other numbers. But they still will never be even numbers, try as they might.

This is not just me talking—just about everybody I've asked feels much the same way once they care to examine it, though curiously, some may even have odd numbers as their favorite numbers. At the very least, they acknowledge that they have different feelings about different numbers, as illogical as it might seem. The only odd numbers that you might risk letting into your exclusive Evens Club—as long as they were properly attired in suit, tie, and shiny shoes—are numbers ending in 5, as for some reason 25, 75, 125, etc., can almost pass as respectable if seen in the right light.

I first became aware of this phenomenon when at the tender age of 18 it was my misfortune to be very badly stoned while in Agra, India, on what I thought was merely a strange-smelling tobacco. It was while in this awful haze that I first recognized the principle that thought is not literal but pictorial. I was amazed to examine at close quarters all those pictures that had been in my head for years and those that popped into my consciousness every time I thought of a word or figure.

So, while strolling through the art gallery of my mind, I very clearly remarked upon the fact that all the pictures of even numbers were Rembrandt masterpieces of smooth-flowing colors melding into one another, while in contrast all the odd numbers looked like they were done by Jackson Pollock on a bad day when all the gray, black, and dark colors had been on sale at Wal-Mart the day before.

THE SEXIST THEORY OF BITCHES AND BULLS

Any animal adjective applied to women is generally insulting, while any animal adjective applied to men is generally directly or indirectly complimentary.

For example, a woman is a bitch, a man is an old dog or a young dog; a woman is a cow, a man is a bull or has bull-like strength; a woman is a cat or catty, a man is a bit of a tomcat (sexual prowess); if a woman has a face like a horse she is ugly, if a man has a horsey face he is pleasantly ugly; and so on ad infinitum. The only apparent exception is a foxy woman and an old fox.

FANCY SEEING YOU HERE!

The warmth of greeting between two people is in direct proportion to the distance between where they now meet and the place where they usually meet.

We work together. I pass you in the office, there is barely any greeting. I meet you in the hall outside, I nod my head in acknowledgment. On the pavement outside we lift languid hands of recognition. I run into you in another state, it's all hail-fellow-well-met! And in Rio de Janeiro we would, at the very least, end up having dinner together after a bare minimum of backslapping, and maybe even bear hugs, if not actually getting married.

PHONE

The lonelier you are, the more likely you are to experience what is known as the "phantom vibration syndrome" on your cell phone.

What is the latter, you ask? Oh, please! As if you don't know! It is that weird sense you get sometimes that your cell phone has just gone off in your pocket, and yet when you haul it out to find out just which lovely person, or close friend, or new friend, or important business contact, or lottery office it is that has called you . . . there is nothing. No one. Just the dead phone. And you.

Dum-de dum dum.

A tumbleweed rolls past. In the distance, a sick horse is heard neighing pathetically, even as a coven of crows gathers on the branches of a dead tree, ready to come down and pick the horse's eyes out when it finally drops. Let's face it—you're as lonely as a stray dog in a cemetery on a windy day. Are you sure that phone didn't just ring? Yup. And the lonelier you are, the more likely it is you'll make the mistake and think it just rang. Don't ask me how I know, I just do . . .

THE WEATHER

The most accurate weather prediction of all is to simply say "the weather tomorrow will be the same as today."

Somewhere or other and awhile back, I read about a whiz-bang weather satellite that had been launched in Canada that was guaranteed to generate weather predictions that were 67 percent accurate. Which was fine, except it was also worked out that simply by invoking the formula above, they were batting 71 percent accuracy!

KNOW-IT-ALLS

Bertrand Russell once said, "The trouble with the world is that the stupid are cocksure and the intelligent full of doubt." I think there is merit in that argument . . . although, on the other hand, of course, you know . . . I can't be absolutely sure.

I can, however, assert that in the world of talk radio, where I have worked on and off for much of the past decade—with mixed success—certainty sells. Across the western world, the highest-rated shock jocks are those for whom every issue is absolutely black and white, with no shades of gray. These jocks know exactly what the problem is and precisely what needs to be done about it! For many, that certainty is very comforting, though I can't resist repeating the theory of Australia's renowned investigative journalist Chris Masters, who reckons that in the realms of talk radio: "The amount of certainty expressed by the media on any topic is in directly inverse proportion to the amount they know about the subject."

TALKBACK

Although there are exceptions, the person who will more often than not rise to the top of the talk-radio tree can fairly be described as the lowest common dominator. That is, a person who is particularly skilled at expressing the most popular prejudices of the populace, in the loudest possible voice, while simultaneously tearing down all those who hold alternative views. All too often, their aim is not to elucidate the truth of any given subject or debate. Rather, they will first give the audience someone to blame for the ills of the day, and then delight in crucifying that someone for days and even weeks on end.

THE GRUMPY-GRUNT THEORY

As a general rule, infantry soldiers start their careers grumpy and get grumpier as the years go by.

This first came to me as I was writing the biography of the former opposition leader, Kim Beazley. Beazley's academic training at both the University of Western Australia and Oxford University had been in the realms of the military and he was confident. Shortly after Beazley became defense minister, he was proud to escort the prime minister, Bob Hawke, along the line of an honor guard of Australian soldiers at their barracks. The soldiers had their entire kit laid out on the ground before them and Hawke, being the chatty sort of man he was, wanted to engage some of them in conversation. In the middle of it all was a weak sort of guy, looking straight ahead in the military manner, but still not managing to suppress a scowl.

"G'day," said Hawke, not at all perturbed. "This stuff looks pretty good. Does it taste all right?"

"No," replied the soldier with some feeling. "It's shit. All of it's shit, and all of it always has been shit, whatever new stuff they give us, it's always *shit!*"

Later, Hawke, still clearly disquieted by what he had heard, and wondering if such discontent in the armed forces might be part of some deep-seated problem, asked the defense minister about it. Beazley put his mind at ease.

"They're infantrymen, Bob," he told him, "what some people call 'grunts,' and that's just the way they are. The thing is, just about every advance in weaponry and warfare techniques over the last 2,000 years has been singularly devoted to more effectively wiping them out, and they're not happy about it. They never will be. Don't let it worry you . . ."

GROUNDHOG DAY

You've seen the film *Groundhog Day*, and so have I. It turns on a tradition in a place called Punxsutawney, Pennsylvania, where if the groundhog ("Punxsutawney Phil") who comes out of his burrow on Gobbler's Knob on February 2 every year sees his shadow, there will be six more weeks of winter weather. If he doesn't, there will be an early spring. All good fun.

There is, however, another theory that purports to tell you what kind of winter it will be, well before that winter has begun. This theory has it that when a tree sheds its bark early in the autumn, it will be a mild winter, while the longer it holds on to it, the colder the winter will be.

GOOSE-NECKING

They call it "goose-necking"—that curious backward and forward movement of the head that people make when they are getting into a rock song but lack the space to really move the rest of their body around. One concert promoter swears by the theory that if people goose-neck during the first song of the show, then the whole thing will turn into a ripper. Alas, if they don't, it's all over, Red Rover.

HI THERE!

When you are walking along and someone you know is walking toward you, then the distance between you and the other person when you say "hello" is in direct proportion to how well you know the person.

If it is a close friend, you will greet each other as soon as you see each other, but if it is just someone you have met once or twice (or just don't know very well), then you will both look somewhere else (anywhere else) until you are within two or three yards and then say hello.

PUBLIC RESTROOMS

In all countries, in all situations, men's restrooms will always be placed in better and more accessible positions than women's restrooms.

More often than not, it is men who design buildings and men who build them. So the "men's" will invariably be in the more convenient location.

Make a note next time you're down at the beach, in a hotel, or pulled over at a rest stop—men's restrooms are, nine times out of ten, the first you'll come to. The women's will be around the back, over wet grass, around the corner or three doors down. The strike rate of success of this theory gets progressively higher the older the building is, but it is still well above the required 51 percent rate even in comparatively modern buildings.

And another thing, while I'm on the subject and getting all these feminist brownie points—despite the logistical inconsistencies, note that men's restrooms and women's restrooms will invariably be much the same size, but the line outside the women's restrooms at busy venues will always be longer . . . for obvious reasons.

GROUP PHOTOGRAPHERS

Photographers say that the "malleability" of a large group of people, in terms of getting them just right for a group photograph, can be roughly judged by their proximity to the age of sixteen.

Getting the Manly Under Sevens football team to line up straight, not fidget, and all smile at one time can be akin to untangling and straightening out particularly sloppy spaghetti, but at least the Under Eights are not quite so difficult . . . and the Under Nines are fractionally easier again.

So it goes, right up to the age of sixteen, when the boys are at their most self-conscious (what with pimples, shaving, approaching manhood, etc.) and they are, consequently, at their most malleable.

A group of seventeen-year-olds, however, will be marginally more inclined to assert their manhood by showing passive resistance to the photographer and on it goes right up to the age of thirty when the *men* really feel they are doing a favor to the little goose of a photographer in the first place, so why not cause havoc? It's apparently somewhere around the age of forty that the group will hit sixteen-year-old malleability again and from there on in, it remains about steady till death . . . when they're the most malleable of all.

ELLIPTICAL RUMORS

The more vicious the rumor, the farther it will travel and the more elliptical its progress will be.

A nice rumor like "Did you hear that Johnny Bloggs is going to marry Sylvia next summer?" travels outward from its source totally unimpeded and produces the same pattern as a pebble when it splashes in a calm pond—nice concentric circles gradually undulating outward, and losing energy all the way.

Truly vicious rumors, on the other hand—"Johnny Bloggs is in fact a cross-dressing bankrupt and won't Sylvia be surprised when she finds out"—follow a far different course. Because it is so vicious, people particularly close to the unfortunate Johnny will either not hear it at all, or hear it and definitely not repeat it. Either way, they stand as impediments to the rumor spreading out concentrically as before.

But in those avenues where the rumor is allowed to move, its very viciousness gives it the energy to travel to the very outer reaches of the subject's acquaintance circle—and sometimes far beyond if it's truly vicious. Sadly, if the rumor is vicious and totally false it achieves time-warp speed and there is little that can be done to stop it, as Winston Churchill knew: "A lie gets halfway around the world before the truth puts its pants on."

THE MAHATMA GANDHI GOODNESS YARN . . .

Less a theory than a good yarn, there is nevertheless a whole philosophy of life contained herein . . .

One day Mahatma Gandhi was getting on a train in Calcutta, surrounded by many of his followers, when in all the hustle and bustle his shoe fell down into the small gap between the train and the platform just as the train was moving off. Without hesitation, Gandhi took off his other shoe and threw it after the first, as his followers looked at him, stupefied. Gandhi explained one shoe was no good to him, and no good to the person who would find it on the track. So why not throw the other shoe down on the track, so at least the finder could have two shoes?

AGAINST THE "DON'T WALK" SIGNS

People are always more inclined to disobey a DON'T WALK sign when a well-dressed person leads the charge. Scrubbers need not apply.

I know this one from experience. I have always totally disregarded pedestrian signals and walked straight across. When I was a scruffy college student, my crossing would have no effect on those waiting. But now that I'm a middle-aged suit, I only have to dip my toe out onto the tar, to test the waters, as it were, before the whole pack bursts forth and carries me along with it.

This rule does not apply in Germany. There, you can be dressed any way you like, and if you disobey the pedestrian sign, a whole pack of them will follow you anyway. But not, heaven forbid, because they intend to break the law . . . au contraire, *mein herr*. They follow because it is already unthinkable for people to disobey the signs, and if someone charges out, many will assume that this is because the lights have changed and they will thus automatically follow without bothering to glance up.

Nor does the rule apply at all in Japan, any time, anywhere, under any circumstances. There, you could put a gun to their heads and they would still not break the law.

The art of concentrating strength at one point, forcing a breakthrough, rolling up and securing the flanks on either side, and then penetrating like lightning deep into his rear, before the enemy has time to react.

—Erwin Rommel, on blitzkrieg

CONSPIRACY THEORIES

The greater the stature of the celebrity, the younger they die, and the more unexpected that death, the more inclination there is to believe conspiracy theories that purport to explain it.

The great John F. Kennedy, the most powerful and celebrated man in the world, shot dead by a poor sap loser like Lee Harvey Oswald? Impossible! It revolts against JFK's very omnipotence. So let us outright refuse to believe it. Let us instead embrace every crackpot theory that comes along: from saying it was the CIA whodunnit, or the KGB, or the Mafia, or the Cubans, or a combination of all four—and let us all be so obsessed by each and all of these theories that an entire industry of publishing and documentaries can be sustained for decades to come. Let that industry thrive on the single notion: It is against both nature and the cult of celebrity that one so high as JFK could be brought down by one so low as Lee Harvey O.

Marilyn Monroe killed by a simple drug overdose? Such a banal ending to such an extraordinary life simply doesn't fit. It is much more fascinating to believe that she was murdered, and even more fascinating if we say it was organized by one or more of the Kennedy family!

Harold Holt couldn't have just drowned because he was so crazy as to go swimming on his own in a wild surf, could he? No, I swear he must have staged his own disappearance, or perhaps been taken by a Chinese submarine.

And all of the above goes double for Diana. What is more interesting? That she died in a traffic accident while sitting without a seat belt in the back of a car being driven by a driver who was so drunk his back teeth were floating? Or that she was killed by MI5 on the orders of the British royal family, because she was contemplating marrying a

Muslim? And did I mention that she was pregnant? And that she'd become engaged to Dodi that very day? And that she had once said to a friend that she feared Prince Charles might one day have her murdered, perhaps by faking a traffic accident?

And so it goes, with Elvis Presley another case in point.

Next time you see a major star on the world stage die unexpectedly young, watch what happens. After the first wave of international grief, look for the second wave—conspiracy theories. The higher the first wave, the broader and more sustained the second will be.

TV SERIES

There is an internal dynamic in TV series that pushes them from the middle of credibility toward the edges as the series progresses—usually with ever-increasing rapidity until eventually the series hits the wall and self-destructs. The usual scenario of successful TV series, such as *M*A*S*H, Dallas, Happy Days, Will & Grace,* and *Cheers,* is that they initially present a roughly believable situation to get the television audience involved. The plot for each episode is drawn from things likely to happen to people in just such a situation. In life there are only so many interesting things that can happen to a person—so too in TV series. Richie Cunningham falling in love with a girl, only to be ultimately rejected, can happen in only three or four episodes before it starts to pall. Two or three years into the series, all the middle-ground territory is used up and the writers have to move outward on the credibility spectrum.

As a matter of fact, the only way to move the series forward is to move it outward, with ever more drastic leaps, until very soonish it happens—the celebrated moment—when the show "jumps the shark"! We label the phenomenon after the famous episode in *Happy Days* when Fonzie—on water skis behind a boat piloted by Richie Cunningham—jumps over a cage in which a hungry shark is waiting.

The show died very shortly thereafter. There's only so far it could go in terms of straining credibility while staying in touch with the show's initial core success, and once it was broken, it was broken for good. The end.

INSURANCE

Insurance salespeople say they never do such good business as when winter first sets in.

In days of yore, as days got shorter and nights got colder, our instincts would switch to self-preservation 'n' safety modes and we would hoard food and wood in our caves to get us through the grim, dark days. These days, we have reasonable expectations that the grocery stores will remain open and that electricity will still get through to our homes, so there is no need to stock the caves. But the instinct of safety-first preparation for the future is still there, and it finds one of its major outlets in buying extra insurance.

> Pray, v. To ask the laws of the universe to be annulled on behalf of a single petitioner confessedly unworthy.
>
> **—Ambrose Bierce, *The Devil's Dictionary*, 1906**

AVIATION

All serious nations claim to boast the real inventor who was crucial in getting aviation off the ground.

We have Orville and Wilbur Wright, while Australia has Lawrence Hargrave, who is credited with developing the idea of curved wings.

But that's just the beginning. China has the man who launched the first hot-air balloon; England, a monk who flew a glider from an abbey just after the turn of the first millennium; Turkey, another glider pilot in 1930; France, the first human in a hot-air balloon on October 15, 1783; Norway, a man who reportedly flew in a glider in 1825; Poland, a fellow who controlled a glider by twisting the wing's trailing edge via strings attached to stirrups at his feet. Phew! Meanwhile, Russia claims to have launched the first "manned multi-engine (steam) fixed-wing aircraft" twenty to thirty yards from a down-sloped ramp as early as 1884, and Germany says its own Otto Lilienthal was the true pioneer of human aviation, as he was the first one to make repeated glides in which he controlled direction.

"That's nothing," says New Zealand, as its own Richard Pearse had witnesses to the fact that he made powered flights on March 31, 1903—including one of over thirty yards—some nine months before Orville and Wilbur Wright made their flights. "Impostors!" claim the Brazilians. Their own Alberto Santos-Dumont's flights three years later brought together everything that had been learned to that point and really established the flying forms on which aviation depends today. And so it goes.

The long and the short of it is, if you can't claim an aviation pioneer in your national history, it is nothing less than humiliating.

True terror is to wake up one morning and discover that your high school class is running the country.

—Kurt Vonnegut

CHAPTER 7

TRAVEL THEORIES

THIS IS YOUR CAPTAIN SPEAKING

Stupendous sights pointed out by "this-is-your-captain-speaking" messages in the middle of long international flights, invariably appear on the left side of the plane.

For the simple reason that pilots, too, never tire of seeing the Grand Canyon . . .

Given that the captain always sits on the left-hand side of the plane, he will always steer so as to give himself the best view. Obviously, the more astounding the sight the more often this theory works.

Thus, if given the choice, you should be among those seated on the left-hand side of the aircraft. But if you hear the words "Afternoon, folks, this is your first officer speakin'," get to the right because you will know that the Grand Canyon, the Hoover Dam, Death Valley, Yellowstone National Park, and the grassy knoll outside the Dallas Book Depository will this time all be on the right side of the aircraft—the side where the first officer sits.

THE AFRICAN SENSE OF TIME

The African people have no sense of time, as we understand it in the western sense.

I once waited for five days in the northern reaches of a desert in the Sudan with sixty totally unperturbed Sudanese students for a train to Khartoum that was marked down in my wretched timetable as leaving at 3 P.M. sharp on Tuesday. Did the train driver have any apologies when he got there? Not a word. Did the Sudanese students give a damn anyway? No. As it happened, they were not blessed with watches, but even if they had been, I'm sure the classic western action of looking at one's watch and furiously tapping the ground with one's foot would have been quite beyond them.

Maybe the train's late arrival *was* wasting time, but hey, they had a whole lifetime's supply of time anyway, so what was the problem? Besides, it was only all the more time to give me a really good thrashing at chess.

It was only after nine days with them (it took four days to get to Khartoum on what was meant to be an eight-hour trip, but that's another story) that it gradually began to penetrate into my western middle-class brain that while we in the west choose to live in an industrial continuum divided up neatly into seconds, minutes, hours, and days, these people divided their lives up into seasons; while if the train hadn't arrived by next winter they might have become a bit huffy, they certainly weren't going to get excited by something so trivial as a few hours or days.

All through Africa I found the same total disregard for the western notion of time, and by the time I got to Johannesburg I had to consciously remind myself that it was the big hand that told you minutes and the little hand that told you hours.

THE TRAFFIC-LIGHT THEORY OF TRAVEL

Take a line of thirty cars waiting at a red light. When the light turns green, car number 1 will move off, followed a second or two later by car number 2, then number 3 and so on . . . until car number 30 begins to move off a minute or so after car number 1.

In Italy, a line of thirty cars will all move off together from the traffic lights, and God help the poor foreign git who doesn't know the rules, as he will be deafened by the blares of horns—if lucky enough not to be smashed into by the cars behind.

The universally recognized definition of a split second should be that period of time on a Rome street between when the light goes green and the bloke behind you honks his horn.

HAND-CLAPPING

There is no better way to judge the rigidity or conservativeness of a society than observing the way its citizens applaud.

Have a look at a shot of a fired-up crowd the next time it comes onscreen. For twenty people who come into view you'll see twenty ways of applauding, from clenched fists and whistles, to wild clapping, to a bit of whatever. Australians are much the same, though perhaps marginally more restrained.

Now look at shots of a Russian or Chinese or even a Swedish crowd. Generally, the twenty people will all be applauding in the same rigid way, one hand stable as the other rigidly comes down upon it.

IT'S MINE!

Flight attendants have noted that when men and women seat themselves side by side in an aircraft, the man will have claimed the mutual armrest within the first thirty seconds.

SLOW TALKING

Worldwide, the flatter the countryside, the slower the speech of its inhabitants.

From the pampas of Argentina to the plains of Spain, from the red flats of Texas to the outback of Australia, the flat countryside seems to make . . . for . . . slower . . . speech.

Why? My guess is that it's because plains are usually far away from the more populated and hilly coastal regions, and the flatter it is, the better suited it is for agriculture on a large scale, and the larger the agricultural scale the fewer people there will be, and the fewer people there will be, the less there will be to say and less opportunity to say it, and the less there is to say and the fewer there are to say it to, the more time you have to say what you're going to say, and the less opportunity you will have to practice your speech and become fast. Phew!

CHAPTER 8

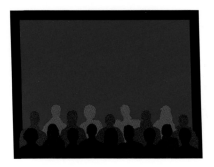

OBSERVATIONS

THE WORLD'S ALL-TIME GREAT OBSERVATION

When in self-serve gas stations, there is a clear difference between the way men and women finish the task of filling their tanks. This is a bit delicate, but it goes as follows . . .

When men have got their required amount of gasoline, they *always* give the nozzle a bit of a shake to extract the last few drops, while the women, on the other hand, only rarely do. Upon questioning, most women are not even aware of the concept.

Now think hard about this: Why should this so very clearly be so? (A friend of mine even goes so far as to maintain that he has twice seen women use a tissue to get

71

the last drops off their nozzles, but I am almost certain he is telling fibs.)

DECORATORS

The cardinal rule for interior decorating is this: Never mix colors that you wouldn't eat together. Because they eventually will make you sick. Examples are chocolate and mustard, avocado and orange, lemon and creamy mashed potato.

JEWELERS

Jewelers maintain that in the week immediately preceding federal or state elections there is always a marked decrease in sales of engagement rings. The more uncertain the election result, the more noticeable this phenomenon. The hypothesis rests on the idea that in uncertain times, really important decisions are delayed until stability returns.

ITCHY NOSE

When you put both your hands into water, your nose automatically begins to itch.

I know this probably doesn't ring a bell with you right away . . . but think again.

The next time you see someone doing the dishes, count the time elapsed between when they first put their hands in the water and when they first use the back of their wrist to scratch their nose. It should be anywhere between one and ten seconds. If this is not the case, be sure to check their pulse, because it just might be that they've died and forgotten to fall over.

DUSTY MOUTHS

Unfailingly, people get specks of dust out of their eyes with their mouths wide open. Always, without fail.

PHONE CALL

Right-handed people tend to hold the phone in their right hand for friendly conversations and the reverse if it is businesslike.

A sudden outburst of anger in the middle of an otherwise friendly conversation often means the person will switch to the left. The reverse is true for left-handed people.

TOOTHPASTE

Planned and organized administrative types will always carefully roll their toothpaste up from the bottom of the tube, while impulsive, creative, damn-the-torpedoes full-speed-ahead types will always take the easiest course and just give the tube a squeeze in the middle and *to hell with the consequences.*

The theory runs that the impulsive ones tend to want what they want when they want it, which is to say immediately, and they don't think ahead to all the problems that will beset them in the future, while the organizers are sure to be more careful about it. They can be seen carefully rolling up the tube from the bottom to be sure to extract the last drops of toothpaste. They may even live for weeks off a tube that the impulsive ones have long since discarded as finished.

A businessman friend of mine once confided that with his executives he always tried to put a squeezer and a roller together on any particular project because, between them, they would be more likely to have the attributes necessary to see the project through with the appropriate amounts of *oomph!* mixed with *steady as she goes.*

The television business is a cruel and shallow money trench, a long plastic hallway where thieves and pimps run free and good men die like dogs. There is also a negative side . . .

—Dr. Hunter S. Thompson

FLOOR WAX

Commercial lemon juice tends to be made from artificial ingredients, while floor wax tends to be made from real lemon juice (or so the blurbs go).

> *The reasonable man adapts himself to the world. The unreasonable man persists in trying to adapt the world to himself. Therefore, all progress depends on the unreasonable man.*
>
> **—George Bernard Shaw**

THE LEGACY OF THE "ME GENERATION"?

There is a clear difference between the way the older and younger generations look at the whole idea of consumer services and goods.

The older generation is still enamored of the qualities of good old-fashioned consumer service with a human touch, simultaneously holding a techno-fear of common domestic appliances, such as computers and DVDs, not to mention the Internet. The current generation, on the other hand, feels more comfortable with machines than it does with people.

Notice how many people would rather wait in line for an ATM in the winter cold than go in and face a bank teller in a short line. The same applies to electronic tolls, where you can frequently see a line before the e-lane while the real, live toll collector sits there practically untroubled.

THE CHECK-VERSUS-CROSS PHENOMENON

When we the people are faced with a little box to be marked, and when there is no firm instruction to check or cross, an amazing pattern occurs.

It doesn't make any difference whether the box is to register preferences for jam or Amex, or 8:30 A.M., or Economy, or to take advantage of our six-months-free offer, the results are the same.

Those in the following occupations or personality groupings check the box: optimists, salespeople, PR people, public figures, politicians, actors, entrepreneurs, and so on.

Those who mark the box with a cross include: managing directors, auditors, actuaries, accountants, professional public servants, those working in large bureaucracies, pessimists, etc.

FAT OR THIN?

When a position that carries some measure of authority is being sought by two people who are exactly equal in all respects, the taller one will be appointed.

If they are the same height, the thinner one will get the job.

However, if the position is a purely support one, such as a secretary, clerk, driver, stenographer, etc., the shorter one gets the job. But the weight preference stays the same—that is, the thinner person always wins if there is nothing else in it.

We must respect the other fellow's religion, but only in the sense and to the extent that we respect his theory that his wife is beautiful and his children are smart.

—Henry Mencken

HELLO?

The people who answer their telephone with their name or company and follow it with "Can I help you?" never help you. They say this to set up a barrier between you both until they find out what you want. Then they set about demolishing you and your request. As a matter of fact, this fits with "Lyndon's observation" on how to define a bureaucracy, which is as follows: If the first person who answers the phone cannot answer your question, it's a bureaucracy.

SO THERE I WAS

After any tragedy, accident, disaster, siege, or disturbance, or just about any newsworthy incident whatsoever, the eyewitnesses always start off their accounts with this: "Well, first I heard a bang . . ."

AND THERE *SHE* WAS . . .

When the time comes to talk about all successful relationships, the happy couple always mention that on the first night they really got to know each other, "We talked until 3 A.M.!"

WHERE HAVE ALL THE NAMES GONE?

Have you ever noticed that in correspondence, men always have at least two names, whereas women often have only one? Look at the bottom of the next few invitations you receive. They often finish with something like this: "RSVP Freddie Flintrock or Susie 212-555-2376" or maybe just "RSVP to Susie."

In a situation where the senior person is female and the junior is male, the invitation finishes: "RSVP Mary Jeffries or Andrew Smith 212-555-2376," never just "Andrew."

The worst situation I ever heard of was reprinted in a magazine. It read: "RSVP: Gareth Wilding-Forbes or Candy."

True.

BRITISH ROCK

British rock guitarists hold their guitar quite clearly in front of their pelvis, thus obscuring their nether regions from view, while American rock guitarists always set their straps so the guitar stays on their tummy, thus leaving their pelvis in full view. Australian rock guitarists fall somewhere in between.

BOYS AND GIRLS

This sounds sexist but it is true nevertheless. When you throw a ball at a girl without warning she will not catch it, but instead bring her arms up to protect herself. A boy, on the contrary, will on most occasions automatically attempt to catch it.

CHAPTER 9

HIGH FINANCE

GOOD LETTERS

A good business letter is about one and a half pages long.

In the business world this is a well-known dictum. Anything shorter is probably too lightweight to bother reading, anything longer, too tedious to get through. One and a half pages, though, strikes just the right balance between attention span and weightiness. Another dictum—this time for the e-mail age, when written communications are more common than ever before—has it that if you have something good to say, write it in an e-mail and send it, but if you have something negative, then it's much better to phone it in.

FAMILY PORTRAITS

Then there's the theory about businessmen who show off their family portraits in the office. The essence of it is this: The larger and more expensive the family portraits, the larger the potential rip-off.

If you are sitting in the sales manager's office of Honest Joe's Used Cars or Trusty's Real Estate and you notice a small package of photographs of the sales manager's wife, son, and daughter in a Wal-Mart frame, then it is likely that you will pay only a couple hundred dollars for the privilege of doing business with this firm. And it is likely that this sales manager gives only an occasional quick leer at Lucy in accounts.

But if you see photographs of the wife etc. in an imported alligator-skin frame—each portrait in glorious technicolor with whiter-than-white teeth beaming out of each photograph through a plethora of jewelry—then you are in for a real beating indeed. And this noble person is undoubtedly having it off with Lucy in accounts and is a little friendly with Suzy in bookkeeping, too.

And when you see family portraits surrounded by ivory or mother-of-pearl, and on another wall there's a montage of happy family snaps, just above the crayon drawings of "I love Daddy" pictures, then you know that you will be lucky to get out of the building with a cent.

And you can be sure that this man is unfaithful to both his mistresses too. Ask the punters who trusted all their life savings to honest investment advisers . . .

The worst case ever reported was that of a man who had in his office a portrait, life-size, of his wife. He was a dealer in fine art. When he left for South America with two of the office women, he wasn't badly missed.

HEMLINES

There is a direct relationship between the length of hemlines and the state of the economy.

Talk to anybody in the fashion industry where this line of reasoning is apparently taken as a given. The theory goes that fashion is inextricably linked to the national mood, and the worse the economy gets the more somber the mood becomes and the longer the hemlines get. The reverse is equally true. Probably because the more upbeat the feeling in the air, the sexier and freer things become and thus the hemlines rise.

> *Organization doesn't really accomplish anything. Plans don't accomplish anything, either. Theories of management don't much matter. Endeavors succeed or fail because of the people involved. Only by attracting the best people will you accomplish great deeds.*
>
> **—Colin Powell, former U.S. secretary of state**

MAKING MILLIONS

Australia is the easiest country in the world in which to make a million dollars.

If only people knew it, or cared to try. As a penniless college student, I was hitchhiking outside our farm on the Pacific Highway between Sydney and Newcastle when it so happened I was picked up by a very talkative fellow in a bright, shiny new BMW who, it turned out, was the owner of what is arguably Australia's most famous brothel, A Touch of Class.

In the course of a highly interesting journey, during which I attempted to drain his brain of everything he knew about life from his curious vantage point, he enunciated the above theory and I have always remembered it since and found reason to believe he was right.

The reasons he gave went like this: First of all, strike out all the communist and extreme socialist countries. They have systems in place to prevent you amassing wealth unless you are a corrupt politician, in which case it will still be extremely difficult, not to mention dangerous, when the regime changes.

Also, strike out all the Second and Third World countries where, by definition, there is very little wealth to go round and what there is will be very difficult to amass. However great India's GNP becomes, it will never come to much when divided by nine hundred million people. Though, of course, it's possible to become very rich in India, when you divide nine hundred million by the number of millionaires, you realize you have about a 0.00003 percent chance of being one of them.

Even if you're born into a wealthy First World country like America, it's still very difficult to make that million for this reason—according to the brothel owner, for every one hundred American adults who walk the earth, probably ninety-five will be engaged in or will have engaged in some get-rich-quick scheme or another because just about everybody wants to be disgustingly wealthy. Ergo, when you line up at the start in the million-dollar race it doesn't look like a 100-yard sprint final with only eight of you going for the prize, it looks very much like the beginning of one of the Ironman triathlon starts where there are literally thousands of well-trained people, many of whom are well ahead of you at the start. You've got no chance of breaking that ribbon first.

Other countries have various crowd sizes at their millionaires' starting line, he said, but nowhere are the starters so thin on the ground as in Australia. Here, there are very few people rattling around in a large wealthy continent and, of those, the vast majority are content to live their lives out under the sun. Thus, he argued, if you really have the gumption, there is little preventing you from winning a race that very few people choose to be in. Therefore, Australia is the easiest country in the world to make a million dollars.

BROTHEL LORE

The circumstances of the last theory made for a hell of an interesting trip, as you can imagine, and it was finished off with a guided tour of the premises. The brothel owner threw in one other interesting theory, based on his experience, that I have remembered through the years. According to him (and I have no reason to doubt him), the turnover on his premises tripled as soon as he came up with the idea of parking a big truck in front of the main entrance, thus obscuring it to all but the few who actually entered the premises. This subterfuge added to an already cleverly obscured foyer, but the parked truck, he said, made all the difference. Curious about its possible efficacy elsewhere, he transmitted the information to a friend of his in the same trade in Perth, Australia, and he also reported a massive increase in trade.

His own ruse lasted for four years until the police got wise to it and created such a blizzard of parking tickets that it no longer became worth it, but the principle was valid.

SELLING PROPERTY

When you are selling a few acres of land, remember this most crucial thing: Always be sure to clearly delineate the boundary. That natural, avaricious streak in people makes them want to clearly see what is theirs. It is not enough to vaguely wave at a bit of bush and say, "What I'm offering you is five acres of this land." Five acres is a hard amount to imagine, but if you can see exactly what will be yours, then you will be much more inclined to buy it.

THE TWO WEEKS' PHENOMENON

Two weeks is *always* the time given to you by a person who has not the slightest intention of carrying out your request.

Remember this on your way up. These people never say, "By Friday or two or three days," because it will still be fresh in your mind by then. Nor do they say, "In three months, or the last week in August," because you will either fight them for an earlier date or enter the date in your datebook and call back when the time's up.

But two weeks is the exact time that makes you temporarily take off the heat—because you actually think the job will get done—and it's a long enough time for you to forget about the whole matter.

MONEY THEORY

Half the money in your wallet disappears every three days.

Think about it. Check it out. No matter how thick or thin your wallet, how great or small your expenses, that basic ratio applies.

FINANCIAL RUIN

The likelihood of any corporation going under can be determined easily by looking at the opulence or ordinariness of its foyer.

It's a classic horseshoe theory: At one end of the horseshoe is the tattered foyer. If it's that bad, then the firm will probably go under because it obviously doesn't even have the money to fix something so basic—or worse, the common sense to realize it is necessary. On the other end of the horseshoe, if the foyer is outrageously opulent, with chandeliers, mirrors, and grand pianos, then the firm will also go under—with such an overwhelming commitment to the superficial, it is a very good bet that the profound and deep is being ignored.

The trouble with being poor is that it takes up all your time.

—Willen de Kooning

THE INVERSE PETER PRINCIPLE

We all know of the famous Peter Principle—established by the Canadian academic Dr. Laurence Peter—which asserts basically that everyone is promoted to their own level of incompetence, at which point they will no longer be promoted. But in these trying times, when people are being fired and demoted rather frequently, it is encouraging to know that there are at least a few shreds of safety net beneath us and that there is an inverse Peter Principle.

Everybody is usually *de*moted only to their level of competence. The exception to this is when you are made CEO . . .

Think about it. What happens as you rise through the ranks? At every level, you keep shining, and keep rising, until the point comes when the demands of the job are greater than your ability to meet them. At this point, one of two things happens. If you are CEO of the company, the board usually invites you to bring a couple of big suitcases to a board meeting, where you are fired. You promptly fill both suitcases with massive amounts of money and walk out the door, shortly before a press release is put out, assuring the market that you are resigning because of "personal reasons."

But if you are not yet at the level of CEO, you will usually find that in the regular corporate reshuffling, you are moved first sideways and a little down, until the next reshuffle, where the same happens again. You keep going until once again you find yourself in a position where you can capably meet the demands of the job and are free to sit at pretty much this level as you watch the desk calendar flutter fast forward just like a calendar in an old movie.

THE EYE OF THE OWNER

Bajo el ojo del amo engorda el ganado—Under the eye of the master, the cattle get fat.

I wish I could say I heard this terrific little aphorism from a bearded bald one on top of an Argentinean mountaintop one moonlit night . . . but in fact I got it from a friend over lunch in Sydney. The basic thesis of this gem is that every business or commercial enterprise goes better and gets fatter wherever the eye of the owner is upon it.

Close to the homestead of the Big Boss, the workers will get things done right and the cows will fatten accordingly, while a few paddocks over from the homestead, things become sloppy. With only the owner's system in place and not his eye, the cows inevitably get thinner.

THE 80/20 RULE OF LIFE

Apparently a standard chestnut among the business troops—known as the Pareto Principle—is that 80 percent of the profit comes from 20 percent of their products, or, put another way, 80 percent of output is produced by 20 percent of input. Similarly, 80 percent of their problems comes from 20 percent of their clients.

They reckon that the ones who make it really big are those who can identify and expand the 20 percent that is making all the money, and identify and eliminate the 20 percent of their clients who are causing all the problems.

THE "ROCKS IN THE BUCKET" THEORY OF TIME MANAGEMENT

We're talking time management, and we're talking the famous line from Rudyard Kipling in his poem "If," where he says the following:

> If you can fill the unforgiving minute
> With sixty seconds worth of distance run
> Then yours is the world, and everything that's in it
> And what is more, you will be a man, my son . . .

In this famous and well-loved story, the question put before the college class is how best to fill that minute with sixty seconds of distance run, the hour with sixty minutes, and the day's work with eight hours' worth of concentrated effort. To answer it, the professor puts on the table a bucket—representing time—some big rocks, some small rocks, some sand, and some water—representing big tasks, smaller tasks, tiny chores, and the rest of the bits of time that make up a day. How best to fill the bucket?

The professor puts the big rocks in the bucket and asks the class: Is it full? Of course not. Now he adds the small rocks and shakes the bucket around, as they filter their way through the crevices. Is it full now? Still not! Then he adds the sand and shakes some more, topping it up with still more sand as space appears. Yes, it is getting fuller now, but of course the pièce de résistance is when he soaks the whole thing with water. Now it is full, and he hopes the lesson is learned: In any given time slot you must tackle the big tasks first, fit the smaller tasks around them, and use the leftover time for tiny chores and whatnots. Clear?

Almost. It is at this point that a student from the back of the classroom remonstrates, saying that the eminent professor is forgetting something. Reaching into his bag, the student pulls out a bottle of beer, walks to the front of the room, and carefully pours it into the bucket with the rocks, sand, and water.

"This shows," he says with a broad smile, "that no matter how frantic you might get in the day, there is always time for a quick beer with your friends . . ."

THE J. PAUL GETTY THEORY OF THE DIVISION OF WEALTH

Undoubtedly the most famous aphorism of the oil billionaire John Paul Getty was this: "Formula for success: Rise early, work hard, strike oil." And yet he also once posited a very interesting theory: "If all the wealth in all the world was evenly divided, and everyone had to start again, then within two years everything would be back the way it was."

See, his view was that by the end of the first couple of days, some people would have either drunk or gambled their wealth away, others would simply sit on theirs and do nothing at all, still others would make questionable investments with no chance of succeeding, and that would mean . . .

Well, that would mean that smarties like him, actually, would inevitably rise in wealth, never giving a sucker an even break, happy to part fools from their money, and making sure that their investments were only calculated risks, rather than foolhardy ones.

I know, I know, in terms of self-serving quotes—à la the famous notion that "some people are born on third base and think they hit a triple"—this one is right up there with the best of them, but I include it on the grounds that it is too interesting not to repeat.

READING

I think the credit for this theory goes to Philip Adams, of radio, authorial, and journalistic fame, and it is very simple. His view is that if you show me someone who reads a book a week, I will show you a good person. And it is true! For almost two decades now, I have been around book people—authors, readers, editors, publishers, and booksellers—and they are, without exception, good eggs. Why? Because if you read a wide variety and a great number of books, it already bespeaks a certain intellect and desire to understand other points of view. It also means that you are exposed to a wide variety of ideas and it is simply impossible to be widely read and to sustain a mean spirit with a closed mind.

COROLLARY

A corollary theory to the last is that TV does not stand for "television" at all, as most people think. Instead, it stands for "time vaporizer." Think about it. Isn't it true that when you turn on the TV at 6 P.M., before you know it, it's 10 P.M. and you can barely remember a single thing you have seen?

The following morning someone could put a gun to your head and you would be incapable of remembering anything! It simply chews up the time, in admittedly a fairly pleasurable fashion, but at the end of it you have absolutely nothing to show for it.

The art of taxation consists in so plucking the goose as to obtain the largest possible amount of feathers with the smallest possible amount of hissing.

—Jean-Baptiste Colbert, minister of finance to Louis XIV

CHAPTER 10

ROCK-SOLID FACTS

Time for another break. Amid all this ephemeral talk that admittedly doesn't amount to a hill of beans, here are a few rock-solid facts to which you can cling while the whirl of all the other theories goes rushing past you like flood-waters at midnight. Everybody should know the following facts, but few actually do . . .

WHITE COFFEE

Coffee with milk tastes infinitely better and creamier when you put the milk in before the boiling water.

No arguments, no discussions. Try it once and you will be hooked for life. This also resolves ever having to worry about the other great coffee/milk question: When you have just made a cup of coffee and the doorbell rings, is it better to put the milk in immediately or when you get back, in terms of its retaining its heat? The answer is immediately, because it will lower the difference between the coffee temperature and room temperature, allowing the coffee to cool at a slower rate. Thus, by putting the milk in first, you'll not only get a creamier taste, but you'll also be well equipped should the doorbell or phone ring.

MILK

While the watched kettle never boils, unwatched milk always does.

ONIONS

The one surefire way not to cry when you peel onions is to take a sip of water and keep it in your mouth as you peel. It works.

THE BEST WAY TO LOSE WEIGHT IS . . .

It is a simple measure having to do with your car that is guaranteed to increase your health and decrease pounds—or at least make you a lot lighter than you otherwise would be. From the moment I began doing it (about six years ago), the weight simply dropped off me, and I have more or less kept it off since. Ready? This is what you do: When you see golden arches on the horizon, *don't* put your blinker on.

It is magic, I tell you!

TOILET PAPER ROLLS

If you place the toilet paper roll with the hanging flap hard up against the wall instead of dangling in the air, it is generally twice as economical. This is because, for some reason, people are far less inclined to go *bbbrrrrrrrrrrrrrrtttttttttt* with the roll when the crucial time comes.

In really luxurious hotels you will notice that the flap will always be out and to hell with the expense, while the cheaper hotels are so Scroogy and nasty as to have the flap against the wall.

TESTES

The word "testify" has curious origins. Way back when, at the height of the Roman Empire, men would affirm in Roman courts that a statement was true by swearing on their testicles. This, by God, certainly made the risks of committing perjury a little more consequential.

A WEIRD LITTLE MATH THING OF NO ACCOUNT . . .

The difference between:
    1 squared and 2 squared is 3
    2 squared and 3 squared is 5
    3 squared and 4 squared is 7
    4 squared and 5 squared is 9
    and so on by the next odd number.

AND WHILE I THINK OF IT, ANOTHER MATH THING . . .

If on June 1st you went to the bank, opened an account, and deposited 1 cent, then on the 2nd you deposited 2 cents, on the 3rd day 4 cents, 4th day 8 cents, and every subsequent day doubled the amount you deposited on the previous day, by the first day of July, you would be worth over ten million dollars . . .

Love is like a Rubik's Cube, there are countless numbers of wrong twists and turns, but when you get it right, it looks perfect no matter what way you look at it.

—Brian Cramer

CHAPTER 11

BODY TALK

BEARDS AND MOUSTACHES

Men with beards basically fall into two categories: those who have pretensions to being intellectuals and those who just can't be bothered shaving. Men with moustaches fall into one category: those who have pretensions to attractiveness.

HUMAN ROUNDABOUTS

Ask a right-handed person to spin around. They will invariably spin around in a clockwise direction. Left-handed people spin in a counterclockwise direction.

AND ANOTHER THING . . .

When I was a kid I often wondered how when my brothers and sisters twirled me around and around with a blindfold on, I still knew which way was north, even in a darkened room. The answer, according to the book *The Compass in Your Nose and Other Astonishing Facts About Humans,* is that humans are equipped with a trace amount of iron in the ethmoid bone between their eyes. This serves as a kind of built-in compass. This is how people who are shut in dark rooms or are blindfolded are still able to align themselves with magnetic north.

SNEEZING

The best way to make yourself sneeze when you're right on the edge of it but just can't quite get there is to quickly glance at the sun.

I've not the slightest idea why this works, but it does for most people, most of the time.

PREGNANCY HIGH

A silly superstition no doubt, but the French say that the higher the lump of a woman's pregnancy rides on her tummy, the more likely she is to have a boy. The Italians, on the other hand, maintain that the more pointy the lump, the more likely it is to be a boy, while the more swollen a woman's back is during pregnancy, the more likely it is to be a girl.

DELICATE

Certain women of my acquaintance assure me that this is neither theory nor conjecture so much as a sure and certain fact, namely that women who live together invariably end up menstruating together. Many women will read this and say, "Yeah, so what? Doesn't everybody know that?"

As a matter of fact, *mesdames*, no, not everybody does know that. Very few men have been let in on the secret, for starters.

MENSTRUAL-CYCLE THEORY

A recent scientific study found that a woman finds different male faces attractive depending on what point she is at in her menstrual cycle.

For example, when a woman is ovulating she will prefer a man with rugged, masculine features. However, when she is premenstrual, she prefers a man doused in gasoline and set on fire, with scissors stuck in his eye and a baseball bat shoved up his backside.

WORD THOUGHTS

Would all kids and Methodists please leave the room so we can get into this one? The theory runs along the lines of onomatopoeic words—words like "splash" and "cuckoo," which sound like the things they describe—except that this theory holds that the look of the words can also be important in the word's ultimate popularity and that this phenomenon is most particularly important in the field of so-called rude words.

Take for example the word "BOOB." Could any other combination of letters have such breastlike curves? With the two Os in the middle, it is just perfect visually. As is the word "BUM," which displays rather well the vision of two buttocks through the U and the M being side by side as well as the sideways curve of the B.

It might be stretching it a bit, but the word "PENIS" also has the same effect, with its P illustrating the rough outline of a dangling penis on top of testicles.

Is it just a coincidence that "VULVA" and "VAGINA" should display so prominently the V, which looks more like a woman's pubic triangle than any other letter in the alphabet? Now look at the internationally recognized symbol for love—the heart shape. In fact the symbol looks nothing at all like what a heart really looks like, but it does look like the silhouettes of two breasts pointing downward.

FOLDED ARMS

Well known, no doubt, but the standard chestnut used by body language experts is that if one party has his or her arms folded in negotiations, it signifies an inflexible attitude. Conversely, the people who use expansive gestures and wave their arms around are far more likely to be open to different propositions.

ELEVATORS

In the comings and goings of elevators in big buildings, people will arrange themselves at all times to remain equidistant from one another.

Three people on the twenty-sixth floor will instinctively stand a yard apart. When two more people arrive on the thirty-second, the original three will subtly adjust their feet a few inches either way to accommodate the newcomers and retain the distance.

Equally, the likelihood of people looking each other in the eye while in an elevator is in direct proportion to the distance between them. While three yards away, you can risk at least a quick glance into their eyes. At one yard, just maybe a look is possible, but still highly dubious. And if hard up against a stranger in a crowded elevator, practically nothing short of a gun in the ribs would make you look in his or her eyes.

BY THE LEFT, MARCH!

On entering a social gathering, restaurant, shopping center, amusement park, or pretty much wherever, a person will generally veer to the left.

There is no doubt this phenomenon exists; it's just a question of why. My favorite explanation is that it's because the heart is positioned on the left side of the body and thus must pump blood to the right side with slightly more force, and thus a step taken with the right foot is slightly longer than a step taken with the left. (Which is why people lost in the desert, or blindfolded, go around in circles.)

If you accept the basic premise of the theory you can use it to your advantage. Simply know that the most influential position at a party or meeting is on the left of the room as you look at it from the door—you will meet everyone who arrives without having to move yourself.

CHAPTER 12

FAMILIES

PUNCTUALITY

Chronic punctuality, or lack thereof, skips generations.

Generally, if one of your parents is hopelessly late, you will be punctiliously punctual and vice versa. Your child will then tend to be the opposite, and so it goes.

FAST EATING

As goes the size of a family, so increases the average speed at which its progeny eat. As goes the order in which the progeny are born, so increases this speed.

Take a look the next time you are at a social function where essentially the same dish is served, and remark upon the order of those who finish first, second, third, and so on. I'll bet that the first few finished will be from large families. I'll also bet that the fastest of the fast will be the youngest of their families or very near thereto.

I qualify for both groupings, being the youngest of seven children, and I don't need any psychologist to tell me why the phenomenon exists. When there's only so much mashed potatoes to go around and only so many sausages to be shared, when all your older brothers and sisters are far bigger and stronger than you are anyway, the only possible way to get seconds is to finish your meal as quickly as possible and then either discreetly serve yourself some more, or otherwise . . . (please feel free to go and make yourself a cup of coffee while I finish this sentence, I'll hold till you get back) . . . or otherwise, I say, if the mashed potatoes are sitting right under your mother's guns, the best way is to stare doggedly over your empty plate into your mother's eyes—trying to get into your own eyes that same glossiness you see in the eyes of a newly caught rabbit—and try to *will* her to ask, "Would you like some more, dear?"

Either way, speed in eating is of the essence. And the thing is, you don't lose the fast-eating habit once you're big enough and ugly enough to flick your hoity-toity brothers into a wall if you have a mind to. It stays with you for life.

Nowadays I know I'm going to get at least my fair share, come what may, but I still eat fast enough that my unkind friends say that after I inhale a hamburger they can see the lump slowly move down my throat.

SUNNY CHILDHOOD . . .

This theory goes that if you had a happy childhood, then when you look back on it, you remember sunny days, the soft breeze of summer coming on, and birds twittering past. If you had, however, an unhappy childhood, then you tend to remember wet Wednesday afternoons, trudging home from school with a hole in your shoe, and your wet sock squelching every step of the way . . .

FAMILY SEQUENCE

Each succeeding child is in fact born into a different family. That is, the first child is born as an only child to generally young and doting parents. The second child is born to slightly older and frequently more tired parents, and has an older sibling. While the third child is one of the litter, being raised by older still parents . . . and so it goes until you get to me. I was born the seventh child, to parents the same age as my friends' grandparents, and at one point my blessed mother was so exhausted by trying to raise me that she asked for the help of all of my older siblings. One way or another, we all muddled through, but my experience in my family was entirely different than that my of brother David, thirteen years my senior and raised, effectively, in a different era.

MAMA'S BOY

In criminal investigations, the one person more suspicious than the butler is the grown man still living at home with his mother. In 1996, I was at the Atlanta Olympic Games and just a few blocks away when someone detonated a bomb in Centennial Olympic Park, killing one passer-by and wounding 111 others. The initial hero of the occurrence was a thirty-three-year-old security guard by the name of Richard Jewell, who had been in the park just before 1 A.M. He spotted a suspicious backpack and cleared the immediate area just before it exploded. For two days he was feted, lauded as the embodiment of everything that was great about an American everyman who rose to the occasion when called upon to go above and beyond the call of duty, until . . .

Until it was discovered that he was still living at home with his mother. Within two days the *Atlanta*

Journal-Constitution pinned him as an FBI suspect on its front page—quickly followed by newspapers world-wide, CNN, NBC, and the like—and his life was never the same again. A terrible trial by media ensued, and always—I followed the case closely—the fact that he lived at home with his mother was at least mentioned, if not focused upon. Though there was never a scintilla of evidence, suspicion hovered all over him until 2005, when an antigovernment militant, Eric Rudolph, was convicted of the crime and Jewell was completely exonerated. Too late. Jewell died in late August 2007, still a broken man.

CHAPTER 13

LIFE AND DEATH

LIFE QUOTES

Come, come now, we're all adults here. Who can really say that one time in your life, I mean one lousy time in your life, you haven't lain awake wondering what life is really all about?

So we all understand one another? It's out in the open now? We can all admit that we do it all the time and that there's no point in pretending otherwise? Perhaps, then, you'd be interested in some of the all-time greatest quotes about life. There is a whole mess of writings on the subject, in every book you've ever looked at, variously hidden, but most of them fade from memory soon after you read them. What follows, after only a few more paragraphs of blabbering (promise) are some that tend to linger—at least for *moi*.

To wit . . .

And we have the judges' envelope—drumroll, please—containing the most depressing life quote of all time. And the winner is . . . Leonard Woolf in *The Journey Not the Arrival Matters:*

> Looking back at the age of 88 I see clearly that I achieved practically nothing. The world today and the history of the human anthill during the last 57 years would be exactly the same if I had played ping-pong instead of sitting on committees and writing books and memoranda . . . I must have in a long life ground through between 150,000 and 200,000 hours of perfectly useless work.

Salutary, no? Aw, come on now. Shut that window and come in off that ledge—I'm sure he was only joking. Sure you were, weren't you, Lenny? He was probably under the influence of what Nietzsche referred to as "the melancholy of all things completed," so it doesn't really count.

But this does. "Common sense tells us that our existence is but a brief crack of light between two eternal walls of darkness," wrote Vladimir Nabokov in his book *Speak,*

Memory. You've got to admit, he's onto something there, no? It certainly puts into perspective the fact that you only placed third in last week's big bowling competition, right?

Sure, the religious types may quibble about there being an eternal wall of darkness waiting for us at the end. But what about this from Somerset Maugham, writing in his deeply reflective book *The Summing Up*: "The religions that men have accepted are but blind alleys cut into an impenetrable jungle and none of them leads to the heart of the great mystery."

Perhaps you holy rollers should think on that the next time you're being stalked by doubts in the silent watch of the night, and pretty soon you'll have all the fun of stumbling around in the jungle with the rest of us. Welcome back.

Incidentally, also with us in the jungle are lots of ultrasober types who live their lives by the letter of ten thousand laws—both real and imagined. This is for you troops, from Gustave Flaubert: "That man has missed something who has never left a brothel at sunrise feeling like throwing himself into the river out of pure disgust."

I mean, if we're going to be stumbling around in the jungle, we may as well live it at least a bit to the hilt, right? Just as Walter Scott said:

One hour of life, crowded to the full with glorious action, and filled with noble risks, is worth whole years of those mean observances of paltry decorum.

So loosen up a bit and let's at least enjoy the jungle for what it's worth.

But on to a few of the famous one-liners about life, painted in the usual gray colors of the truly bleak landscape . . .

There is Fran Lebowitz's: "Life is something to do when you can't get to sleep";

R. D. Laing's: "Life is a sexually transmitted disease, and there is a 100 percent mortality rate";

and John Lennon's most famous "Life is what happens to you while you're busy making other plans."

For some reason I haven't yet been able to fathom, just about everything memorable anybody ever said or wrote about life that wasn't encased in a hymn seems to be downbeat rather than upbeat, but maybe they're the only ones that appeal to me. (Razor anyone?)

The most neutral one-liner about life is Mahatma Gandhi's: "There is more to life than merely increasing its pace." But he is reputed to have said this as a twelve-year-old he had just come in second in a one hundred–meter race, so we can't be sure he really meant it. These days, he'd surely be lining up for a cell phone like the rest of us.

One of Bob Dylan's great anthems, specifically about life, was "Mixed Up Confusion," and it is in the grip of just such emotions that some have turned to . . . drugs.

To be fair, though, they have worked for some. As the late, but definitely great, Hunter S. Thompson said: "A cap of good acid costs $5 and for that you can hear the Universal Symphony with God singing solo and the Holy Ghost on drums."

But there was a rider added to Thompson's thesis by William F. Buckley Jr. that with drugs "one should be prepared to vomit rather frequently and disport with pink elephants and assorted grotesqueries while trying, often unsuccessfully, to make one's way to the toilet."

So after all that, are you any closer to the answers? No, me neither, but if you just turn up the violin music a fraction louder, maybe this will make us all feel a bit better . . .

"Life is an unanswered question, but let's still believe in the importance and dignity of the question."

Thank you, Tennessee Williams.

Fade to black.

DEATH QUOTES

So there he was. Albert Einstein, at death's door for so long he was starting to be mistaken for its knocker . . . He stirred. He moaned. He spoke. He died.

What a pity his last words were in German, a language the attendant nurse at Princeton Hospital didn't speak. If we are to believe *Omni* magazine, which recently revealed the incident, it was because of this simple glitch that the genius, whose theories of matter and energy transformed humanity's understanding of the earth and the universe for all time, went to his grave with his final words unrecorded.

Might it have been that his final words were: "Ggjh . . . E does . . . not equal . . . MC squared . . . njnhh . . . after all"? We'll never know.

The propensity of history to record famous people's last words wasn't given much of a fighting chance in this case. Not that that has necessarily stopped it before. Whether Horatio Nelson actually did say, "Kiss me, Hardy," on the poop deck of the *Victory* before he expired, or George V really did say, "Bugger Bognor!" as the last thing before he went under are matters highly questioned by serious historians. In the end it doesn't really matter.

The fact is, it is all but impossible for posterity to record that someone like Oscar Wilde could ever shuffle off this mortal coil without one last burst of wit. In his case, posterity overcompensated by recording two possible farewell lines: "Either this wallpaper goes or I do" and "I am dying as I have lived—beyond my means."

Très, très droll, Oscar, whichever it was. Not his best, but who can blame him? It must have been hard to come up with something really good, feeling as ghastly as he did.

Following, though, are the last words recorded of a few other famous people. There is General John Sedgwick, whose last words at the battle of Spotsylvania in the American Civil War earned him an immortal place in the Tragicomedy Hall of Fame. The good general, in all his magnificent finery, had been prancing up and down behind his rebel soldiers, exhorting them to ever-greater efforts to repel the Yankee invaders.

"Sir," ventured one of his underlings, "don't you think it would be a good idea if you were to present a less obvious target to those damn Yankees?"

"Don't be ridiculous!" boomed the fearless general. "They couldn't hit an elephant at this dist—"

No doubt apocryphal, but who cares? If that isn't what he actually said, that's what he should have said.

Speaking of which, the most famous of all (Australian) farewells was no doubt that of Ned Kelly, who, as he stepped up to the hangman's noose, uttered the rather stoic and magnificently philosophical "Such is life." (Actually, when you think about it, those words are not really magnificently philosophical, but almost anything Kelly had said under such circumstances, that was not a plea for mercy, was bound to be celebrated for its heroism.)

Seventy-odd years later, another outlaw would go to his death with an equally famous farewell on his lips. Neville Heath, a British murderer, was executed in the final months of 1946. Asked if he would like to have a final drink, he said he would rather like to have a whiskey. There was a pause as the governor nodded to an orderly to go get one, and Heath called out after the departing man ". . . and you might as well make that a double." Indeed.

For real magnificence, one need go no further than Captain Lawrence Oates's gesture on his way back from the South Pole with Robert Falcon Scott of the Antarctic. Knowing that his disability would slow the expedition down—to their almost certain death—he committed history's most heroic suicide by heading out into a blizzard in the middle of the night. At the tent flaps he turned to his prone companions—who made not a murmur, moved not a muscle—and said: "I am just going outside and may be some time." And he was gone, never to return.

Sometimes one wonders if famous last words didn't have an effect on later poetry. Goethe's famous *Mehr Licht!* (More light!) goes so very nicely with Dylan Thomas's line of advice to his dying father—"Do not go gentle into that good night . . . Rage, rage against the dying of the light"—

that one can't help but wonder whether Thomas didn't hear of Goethe's last utterance.

But is there an actual dying of the light when the final hour comes? Hard to say . . . In all the famous last words, there is little to be garnered from what the experience—if one can call it that—is actually like, though there is at least some suggestion that it might be something like the beginning of a big roller-coaster ride.

There is famous boxer Max Baer's "Oh, God, here I go!" and Al Jolson's "This is it. I'm going! I'm going!" A more placid departure was that of Thomas Edison, who came briefly out of a coma to say, "It is so very beautiful over there!" (On the other hand, maybe it wasn't so placid after all. In the Henry Ford Museum in Dearborn, Michigan, a corked test tube is purported to be Edison's last breath. Did they shove a test tube into his mouth straight after he said this? It defies belief.)

Then there was Buddha: "Never forget, decay is inherent in all things."

Philosophy is not for everyone, though. Others like a quick, blunt assertion of fact. For sheer irrefutability there is none to beat the famous French writer Dominique Bouhours: "I am about to die." And promptly did.

It's hard to forget Dr. Joseph Green, an English surgeon who died in 1863. While lying on his deathbed, Dr. Green looked up at his own doctor and said, "Congestion." Then he took his own pulse, reported the single word "stopped," and died.

Finally, for the truly touching, there is Alfred Shaw, an ex-England cricket captain, who died in 1907 four years after his best friend and fellow cricket player, Arthur Shrewsbury, who had committed suicide. His final words were "Bury me twenty-two yards from Arthur, so I can send him down a ball now and then."

CHAPTER 14

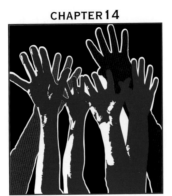

GOVERNMENT AND WORLD POLITICS

THE GOLDEN ARCHES THEORY OF CONFLICT

This theory, posited by Pulitzer Prize–winning journalist Thomas L. Friedman, has it that "there never have been, and likely never will be, armed conflicts between two countries where both have McDonald's."

The theory stands on two pillars. The first is that only countries with substantial middle classes can sustain McDonald's, and the middle class has too much to lose to want to engage in a war. As Friedman puts it, "People in McDonald's countries do not like to fight wars anymore; they prefer to wait in line for burgers—and those leaders or countries that ignore that fact will pay a much higher price than they think."

The second thing, of course, is that having golden arches dotting your landscape bespeaks a certain level of "sophistication" that is likely reflected in the power of your armed forces. This means that whereas most armed conflicts start when one very powerful country invades or bombs a very weak one, having golden arches on the horizon is probably a fair indication that you will be able to provide some sting in return. The only possible exception to this rule in recent times has been the Kosovo conflict, which started as a civil war in what was Yugoslavia and finished as an armed conflict between that country and NATO forces.

Everyone is always in favor of general economy and particular expenditure.

—Anthony Eden

DEMOCRACY

Our political system's two broad parties that are characterized as "Left" and "Right," can also be understood as "Mommy" and "Daddy" parties.

This theory has it that just as kids need the succor of Mom and the firm hand of Dad to grow into balanced adults, so too does a country's population need regular bouts of both parties in power to prosper and grow.

The Mommy Party of the left will be excellent at ensuring that the nation's apple pie—its gross domestic product—is fairly divided. And any children who can't feed themselves will always have Mom on hand to look after them, come what may, no questions asked. If one of the children does the wrong thing, Mom will be understanding and, instead of meting out instant and heavy punishment, will at least give the child a fair warning. Then and only then will Mom hand out limited punishment, if at all. Mom's idea of security in the face of danger, so the popular imagery goes, is to simply hug the children close and let them know they are loved.

Careful, though: "Spare the rod and spoil the child." Too much indulgence from the Mommy Party and it becomes obvious that the more stringent approach of the Daddy Party is needed.

The Daddy Party, see, is less concerned with how the apple pie is divided and is far more focused on ensuring that it is a big apple pie to begin with. And the way Dad sees it, those who had most to do with creating that pie should get the biggest portions, no questions asked. As an example, any kids who say they're too sick or tired to do their chores today have to learn the hard way that life is tough—no apple pie at all for them! (And I'll bet they won't feel so tired or sick tomorrow.)

Now, if Junior crosses the line and, let's say, deliberately breaks a plate while washing up, Dad doesn't have the time or care to work out why he did it. Rather, Dad believes there is no transgression a quick, hard smack can't fix, while still reserving the right to send the child to his room for a long, long time if necessary and . . .

And what was that noise?

Someone, perhaps a burglar, moving around in the darkness outside? Well, that's fine for Mom to hug the children close, but Dad knows that someone has to arm himself with a bat and just head on out to see who's there.

Writ large, you have Mommy governments of the Left promoting welfare and equitable divisions of national spoils, even while they take a generally benign view of miscreants who do the wrong thing. This all works well to a point, until the electorate feels that things have become too soft, at which point they elect a Daddy government as a corrective. The Daddy government is far less keen on welfare than having everyone work for a living, and has no patience for those who break the law. On matters of national security, Dad believes in strength first and foremost. This all goes well for a time, until the electorate inevitably tires of it and elects a Mommy government as a corrective.

And so it goes . . .

The theory also explains, in federated nations like Australia, why as a general rule the states—with responsibilities like health and education—tend to be ruled by the Mommy Party of Labor, while the national government—concerned with national security and the economy—is more likely to have the Daddy Parties of the Coalition in command.

Power tends to corrupt; absolute power corrupts absolutely.

—Lord Acton

POLITICS

The winner of any federal election in Australia can be determined by consulting a very simple formula. In the term of the incumbent government, look to what has happened to rates of interest, inflation, and unemployment. If at least two of the three have gone up since the last polling day, the incumbent government will lose the election. If at least two of the three have fallen, it will win the election.

The only exception is when the term is shorter than two years, in which case the incumbent will win (as happened in 1974 and 1984). The theory was developed by a South Australian, Wattie Sawford—grandfather of the Federal Labor Member for Port Adelaide, Rod Sawford. It has worked in every election since almost forever, including the 1961 election, when the Menzies Coalition government just managed to hold off Labor's Arthur Calwell by the handful of votes that determined just one seat (with Wattie Sawford predicting that very result).

In recent years, Rod Sawford himself has become so enamored of the theory that he has refined it into a mathematical matrix that he says allowed him to predict that George W. Bush would win the 2004 presidential election by about 100,000 votes and that Labor would lose the Mark Latham–led election that same year by the margin he did.

Naturally, the common people don't want a war, but after all it's the leaders of the country who determine the policy, and it's always a simple matter to drag the people along . . . All you have to do is tell them they're being attacked and denounce the pacifists for lack of patriotism. It works in every country.

—Hermann Goering (At His Nuremberg Trial)

LOG-CABIN LEADERS

Social analysts the world over note that the best environment for producing leaders—be it in industry, commerce, politics, agriculture, or whatever—is the small rural township.

Time and again, when you look up the biographies of leaders in *Who's Who* or something, you will see something to the effect of "born and raised in Smallville, moved to the Big Smoke at the age of eighteen."

In their early years, young leaders get a taste for running the show in a stable, supportive environment and learn the principles of good leadership. They also develop the confidence, knocking over small-town problems, to put these principles into effect.

Why specifically rural towns and not elsewhere? Because when you grow up two miles from the black stump and your nearest neighbor is twenty miles away, there are no organizations from which to learn leadership in the first place . . . and if you're born and raised on the mean streets of the city, not only are there a million other distractions other than becoming a Boy Scout leader, but with that nascent city-cynicism in you, you don't get to develop that wholesome down-home appeal that goes so well toward advancing one's cause in politics, business, and whatever.

In recent years, from just the political field, one points to Ronald Reagan, Jimmy Carter, Mikhail Gorbachev, and Golda Meir as obvious examples of the phenomenon.

The less you intend to do about something, the more you have to study it.

—Sir Humphrey Appleby, *Yes Minister*

ROYAL COMMISSIONS

In any royal commission, the level of blame or culpability is inversely proportional to the health of the accused.

That is, *dead* people will be found to have acted most wickedly of all, followed closely by those who are unconscious or on life support systems, then the healthy but aged. Least blameworthy will be the fit and young.

Being powerful is like being a lady. If you have to tell people you are, you aren't.

—Margaret Thatcher

YOU *DO* NEED A WEATHERMAN TO KNOW WHICH WAY THE WIND BLOWS

Elections will nearly always be called in times of the most temperate weather.

The record bears this out. The logic runs thus: The government wants to maintain the status quo. The people will be more inclined to maintain that status quo when things are going well. The weather plays a big part in this collective feeling of the people. A savvy government, therefore, will always call an election at a time when the people feel good. And when is this time precisely? Do we ever feel better than when the weather is nice?

The easiest way to create a radical is to belt a conservative over the head with a police truncheon.

—Keith Suter, Australian academic

CHAPTER 15

INTERNET THEORIES

Since the first edition of this book, the Internet has arrived as a global phenomenon with billions around the world now online. Apart from everything else, it has been wonderfully fertile ground for little theories to circulate. Over the years, I have set the best, or at least the most amusing, aside, and here are a few of them . . .

A RELATIONSHIP THEORY ALL ITS OWN—"JUST FRIENDS"

The basis of this theory is wonderfully expressed in the 1989 film *When Harry Met Sally*. You will recall that when Harry (played by Billy Crystal) is in a car on the way to New York with an acquaintance, Sally (Meg Ryan), they begin to discuss whether a man and a woman can be "just friends."

Harry says it simply isn't possible, because the sex part always gets in the way, while Sally insists that it is quite possible, as she has plenty of male friends with no sex involved. (Remember the scene?)

Wryly amused, Harry insists that while she is not having sex with her male friends, there is no doubt that they would like to have sex with her!

Welcome to what has become known as the Ladder Theory, written by an author who chooses to remain anonymous but whose work can be found simply by Googling the aforementioned theory or simply going to www.laddertheory.com.

Okay, all ready?

THE LADDER THEORY

The theory put up by this anonymous genius, in essence, is that every time a man meets a woman, he instinctively places her on a ladder in his mind that ranks her according to this key question: "Just how much would I like to sleep with this woman, and what are the chances of it happening?"

True, it is a question that might be asked subconsciously, but it is posed all right, and it guides the man's conversation during that first meeting, as he works out where to situate her. At the top is his dream partner to beat them all; at the bottom is what is known as "the Abyss," consisting of women he wouldn't dream of sleeping with. The theory has it that while of course he has women who are just friends, he still ranks them on his ladder, because he knows that if the opportunity arose, he really would sleep with them.

A woman, however—and here is the rub—has two ladders. All males she meets are divided up—subconsciously or otherwise—into those who are prospective sexual partners and those she would like to have as friends. She is not so gauche as to have a "Sex Ladder," of course, but prospective sexual partners fit on her "Good Ladder," while the others are on her "Friends Ladder." Those who don't make it into either, alas, fall into—you knew it all along—the Abyss.

Still with me, tree people? God bless you, I'm singing for you, too.

Okay, given that the man has a sole ladder and the woman operates on a "bi-ladderal" system, what frequently happens is that while the woman will find herself very high on his ladder, he finds that he himself is on her Friends Ladder only! Surely there must be some mistake?

Nooooo, alas not.

What now happens, all too frequently, is that the man will attempt a ladder jump, which very occasionally is successful (see: Once in a Blue Moon Theory) but all mostly end with him falling into—again!—the Abyss.

Now—according to our anonymous genius—many guys are not put off by the number of men who have tried the ladder jump and failed, and so they content themselves for the moment with getting as high on the Friends Ladder as they can before making the jump. Ladder Theory puts this type of man into the category of "intellectual whore": "A man who readily pays a woman with his emotional support in the hope of receiving physical returns."

See, this guy won't front up and make it clear early on that he wants to be on her Good Ladder, to be seen as a prospective sexual partner. No, he'll take the seemingly easy option and climb the Friends Ladder by buying her gifts, being there when she is down, taking her to the theater, and giving her only a peck on the cheek on the doorstep. He knows what he is doing, but she doesn't.

If he's lucky—or particularly pathetic, depending on how you look at it—he might graduate from being a mere IW to being a Cuddle Bitch. He settles for hugs and even spooning—but nothing more—while she pours out her heart about the guy she is having sex with (known as the Outlaw Biker)—the nasty bastard who doesn't emotionally care about her at all and who doesn't bother hiding it- and whom she is therefore interested in!

What to do? Our man, the Cuddle Bitch, can stand this for only so long before he eventually decides to make the ladder jump, figuring he is now so high up the ladder he will make it across the gap. Alas, alas, he fails just the same and falls into the Abyss.

Cue Harry: "The friendship is ultimately doomed, and that is the end of the story."

And this, dear friends, is just the beginning of the Ladder Theory.

THE BUFFALO THEORY OF BEER

We all remember Cliff on the sitcom *Cheers* and his buddy Norm, who were two of the main characters. (Oh come on, of course you do. Cliff was the mustachioed postman who propped up one end of the bar, while Norm was his rotund business buddy who propped up the other.) Anyway, one afternoon at Cheers, Cliff was explaining to Norm the Buffalo Theory of Beer . . .

The essence of it was that since a herd of buffalo can move only as fast as its slowest member, it automatically follows that when the herd is hunted or attacked the slowest ones are inevitably killed, and the herd becomes faster and stronger. Regularly killing off the weaker members, then, makes for the overall health of the herd.

And so Cliff finished with the insight that the human brain works the same way: Excessive alcohol kills brain cells but it naturally attacks slower and weaker cells first. So, says Cliff, "regular consumption of beer eliminates the weaker brain cells, making the brain a faster and more efficient machine. That's why you always feel smarter after a few beers."

Look, it works for me!

THE INTERCONNECTEDNESS OF EVERYTHING THROUGH THE AGES (OR HOW THE SPACE SHUTTLE WAS IN PART DESIGNED BY THE ANCIENT ROMANS)

The international standard railway gauge (distance between the rails) is a smidgen under 4 feet, 8.5 inches. It is a strikingly odd width, but it is, nevertheless, the standard gauge. Why? Because that's the way they built them in Britain, and the Brits built the railways in the United States, Japan, and many other countries around the world as they desperately tried to continue to expand their crumbling empire.

The gauge is a legacy of the earlier tramways; railway engineers used the same tramway jigs and tools for building the railway wagons. The tram coaches, in turn, had been built using the standard jigs used to build horse-drawn stagecoaches. Horse-drawn carriages had that wheel spacing because it was the standard spacing of the wheel ruts gouged in the roads.

Long-distance roads in Britain (and elsewhere in Europe) were first built by the Romans as they expanded their empire. The Romans had earlier experimented with the design of battle chariots and found that two horses, side-by-side, was the best arrangement in terms of speed and maneuverability. Although their engines were only two horsepower, we shouldn't underestimate the importance of these early test pilots. Their findings fixed the width of chariots, and to ensure wheels wouldn't get stuck, roads were designed to accommodate them.

Now let's move forward a few thousand years to the design of the space shuttle. You may have noticed that there are two huge ancillary rockets on the sides of the main fuel tank. These are known as reusable solid rocket boosters and they are made by a killing-machine firm called ATK Thiokol in Utah. The assembly plant for these boosters is on the western side of the Rocky Mountains, and to transport them to the Kennedy Space Center in Florida, rail is the cheapest and quickest method. The rail line from the factory runs through a tunnel, so the booster rockets have to be made smaller than the tunnel. This tunnel is only a bit wider than the track, and therefore the rocket design had to take this into consideration.

So there we have it. Today's space travelers owe the design of the shuttle's rocket motor to ancient Roman test pilots.

THE SIX PHASES OF A PROJECT

1. Enthusiasm
2. Disillusionment
3. Panic
4. Search for the guilty
5. Punishment of the innocent
6. Praise and honors for the nonparticipants

ROMANCE MATHEMATICS

Smart man + smart woman = romance
Smart man + dumb woman = affair
Dumb man + smart woman = marriage
Dumb man + dumb woman = pregnancy

GENERAL EQUATIONS AND STATISTICS

A woman worries about the future until she gets a husband.

A man never worries about the future until he gets a wife.

A successful man is one who makes more money than his wife can spend.

A successful woman is one who can find such a man.

CHAPTER 16

LITERATURE THEORIES

Great literature, too, has passages that contain, in singularly eloquent prose, theories that have lasted the test of time.

A BETTER MOUSETRAP

"If a man has good corn, or wood, or boards, or pigs to sell, or can make better chairs or knives, crucibles, or church organs, than anybody else, you will find a broad, hard-beaten road to his house, tho it be in the woods. And if a man knows the law, people will find it out, tho he live in a pine shanty, and resort to him. And if a man can pipe or sing, so as to wrap the prisoned soul in an elysium; or can paint landscape, and convey into oils and ochers all the enchantments of spring or autumn; or can liberate or intoxicate all people who hear him with delicious songs and verses, 'tis certain that the secret can not be kept: the first witness tells it to a second, and men go by fives and tens and fifties to his door."

Ralph Waldo Emerson

(This is the origin of the famous line: "If a man builds a better mousetrap, the world will beat a path to his door.")

SELF-EVIDENT TRUTHS

"It is a truth universally acknowledged, that a single man in possession of a good fortune must be in want of a wife. However little known the feelings or views of such a man may be on his first entering a neighborhood, this truth is so well fixed in the minds of the surrounding families, that he is considered as the rightful property of some one or other of their daughters."

Jane Austen

(These opening lines of *Pride and Prejudice* are indeed a self-evident truth!)

PERSONAL FINANCES

"He solemnly conjured me . . . to take warning by his fate; and to observe that if a man had twenty pounds a year for his income, and spent nineteen pounds nineteen shillings and sixpence, he would be happy, but that if he spent twenty pounds one he would be miserable."

Charles Dickens's Mr. Micawber instructing David Copperfield on organizing his personal finances

"CUCKOO CLOCK THEORY"—OR WHY A CERTAIN AMOUNT OF DISORGANIZATION CAN ACTUALLY BE GOOD FOR YOU

"In Italy, for thirty years under the Borgias, they had warfare, terror, murder, bloodshed—they produced Michelangelo, Leonardo da Vinci, and the Renaissance. In Switzerland, they had brotherly love, they had five hundred years of democracy and peace, and what did that produce? The cuckoo clock . . ."

Orson Welles

DEEP THOUGHTS

"Every man has his own patch of earth to cultivate. What's important is that he digs deep."

Novelist and Nobel Prize winner Jose Saramago

SHAKESPEARE IS A FRAUD

For years the greatest conspiracy theory in the world of literature has been that the true author of the collected works of William Shakespeare was not in fact Shakespeare himself, but someone else—Christopher Marlowe, Francis Bacon, the Earl of Oxford, Edward de Vere, are all frequently cited as the true authors.

This theory has it that Shakespeare himself—with his poor background, lack of travel, and lack of classical education—could not possibly have written such extraordinarily learned plays, and therefore he must have been a mere front for another who wrote the plays but did not want to claim them as his own.

Why wouldn't that person have claimed such stunning plays? Because, the theory goes, writing plays at that time was not a worthy pursuit for a nobleman classicist, and so one who was capable of it might well have allowed Shakespeare to put his name upon them to hide his own identity. No less than Samuel Taylor Coleridge gave the theory a push along by saying the following: "Ask your own hearts, ask your own common sense, to conceive the possibility of the author of the Plays being the anomalous, the wild, the irregular genius of our daily criticism. What! Are we to have miracles in sport? Does God choose idiots by whom to convey divine truths to man?"

Over the decades, and then centuries, attention has particularly focused on Francis Bacon. A contemporary of Shakespeare, Bacon was a brilliant writer and is regarded as having had the education necessary to write such plays. On that subject it is James M. Barrie who has put it best: "I know not, Sir, whether Bacon wrote the works of Shakespeare, but if he did not it seems to me that he missed the opportunity of his life."

Of course, nothing has ever been proven, and likely nothing ever can be proven, though I have long treasured the story about a computer program that was invented in the early 1970s purporting to "prove"—by analyzing the structure of the language used—that the author of Shakespeare's and Bacon's works were one and the same man. As I remember hearing the story, reporters at the *Chicago Sun-Times* were very excited about this idea when they discovered it and were going to go big on it in the following day's edition, until one of the copyboys ran the computer program over the daily racing guide and proved that Francis Bacon had written that too!

CHAPTER 17

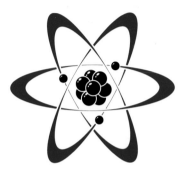

LAWS, THEOREMS, AND DICTUMS

And now, the end is near, and so we face the final curtain . . .

For a finale, however, here are some pithy words so wise that over the decades, and even centuries, they have acquired the sheen of natural laws.

PARKINSON'S LAW I

Work expands so as to fill the time available for its completion.

Cyril Northcote Parkinson

PARKINSON'S LAW II

The man who is denied the opportunity of taking decisions of importance begins to regard as important the decisions he is allowed to take.

Cyril Northcote Parkinson

ERMA BOMBECK'S RULE OF MEDICINE

Never go to a doctor whose office plants have died.

ACHESON'S RULE OF THE BUREAUCRACY

A memorandum is written not to inform the reader but to protect the writer.

BOVE'S THEOREM

The remaining work to finish in order to reach your goal increases as the deadline approaches.

COHN'S LAW

The more time you spend in reporting on what you are doing, the less time you have to do anything. Stability is achieved when you spend all your time reporting on the nothing you are doing.

GREENER'S LAW

Never argue with a man who buys ink by the barrel.

MURPHY'S LAW

If something can go wrong, it will . . .

LORD BEAVERBROOK'S LAW

Organization is the enemy of improvisation.

GOLDEN RULE

The golden rule of the business world is that whoever has the gold gets to make the rules.

THE CRUICKSHANK COMMITTEE THEORY

If a committee is allowed to discuss a bad idea long enough, it will inevitably vote to implement the idea simply because so much work has already been done on it.

In every work of genius we recognize our own rejected thoughts . . .

—Ralph Waldo Emerson

THE CARPENTERS' CREED

Measure twice, cut once.

THE ROCK 'N' ROLL ROADIES RULE

If it moves, root it; if it doesn't, put it in the truck.

BART SIMPSON'S THEORY OF LIFE

I didn't do it. Nobody saw me do it. There's no way you can prove anything.

Thank you, thank you all. And good niiiiiiiight . . . !

About the Author

PETER FITZSIMONS is one of Australia's bestselling authors. He is a well-respected columnist for the *Sydney Morning Herald* and *Sun-Herald*, and a broadcaster on Radio 2UE He is an ex-Wallabies rugby star, a television presenter, and a regular contributor to the *International Herald Tribune* and *The Daily Telegraph*.

ACKNOWLEDGMENTS

As promised, my acknowledgment of those people who helped me put this book together by contributing odd knickknacks and whole theories. I am indebted to them all, but most particularly Jim McPherson of Melbourne, Australia; Suzanne Falkner of Sydney; and Mick Johnson of Merriwa, Australia, whose help was invaluable. As to my brother, James FitzSimons, if the truth be known, three of the best theories are his and our conversations about the final form of the manuscript seemed endless.

Milto Baratella
Richard Bean
Peter Bowers
Helen Chisolm
James Cockington
Stephen Cutler
Dominic Di Biase
Paul Donovan
Neil Dunn
Deborah Edwards
Mark Egan
David Eggleston

Kathy Elrington
Darrell Fazio
Peter Fenton
David FitzSimons
Stephen Freed
Patrick Gallagher
Tony Hernandy
Robert Haupt
Alison James
Mrs. J. H. Jones
Paul Latler
Cheryl Long
Ewen Mackenzie
John Manning
B. J. McKusker
Jane Mulligan
Sian Powell
Peter Roebuck
Penny Sheehan
Peter Skinner
Mr. S. Stebellini
Paul Stokes
Lara Tayles
Dot Thompson
Paddy Thurbon
Tony Villa
Conrad Walters
Col Whelan
Tony Zuccharini